Advanced Chassis Control Technology for Steer-by-Wire Vehicles

Advanced Chassis Control Technology for Steer-by-Wire Vehicles details state-of-the-art drive-by-wire technology, enabling engineers to create safer and smarter steering technology. With applications in Formula 1 driving, this book is an accessible yet ambitious introduction to the technology that is fast becoming the future of road vehicles.

Steer-by-wire systems replace conventional mechanical technology with electronic sensors, controllers, and actuators, enhancing functionality when steering. Features such as variable steer ratio, customized road feel, and advanced vehicle dynamics control all ensure that this maximizes safety when driving. The book looks first at the theory behind this technology and compares it to conventional mechanical steering. It discusses control through forward and backward dynamics and a shared steering control concept to improve vehicle handling and performance, relevant to intelligent vehicles. It also explains how to create chassis domain fusion control, four independent wheels steering system and teleoperated control. Using case studies and ISOs, the book is a practical guide to safely designing steer-by-wire systems.

The book is an essential guide to all engineers working in the modern automotive industry.

T0320687

Advanced Chassis Control Technology for Steer-by-Wire Vehicles

Xiaodong Wu

CRC Press
Taylor & Francis Group
Boca Raton London New York

CRC Press is an imprint of the
Taylor & Francis Group, an **informa** business

First edition published 2024
by CRC Press
2385 NW Executive Center Drive, Suite 320, Boca Raton FL 33431

and by CRC Press
4 Park Square, Milton Park, Abingdon, Oxon, OX14 4RN

CRC Press is an imprint of Taylor & Francis Group, LLC

© 2024 Xiaodong Wu

ISBN: 9781032740775 (hbk)
ISBN: 9781032771793 (pbk)
ISBN: 9781003481669 (ebk)

DOI: 10.1201/9781003481669

Typeset in Times
by Newgen Publishing UK

Contents

Preface

With the development of electronic control technology, by-wire systems, which are derived from aerospace technology, have initiated possible replacement of mechanical and hydraulic parts on these vehicles. One of these advanced control technologies is the steer-by-wire (SBW) system. In the steer-by-wire system, compared with traditional steering systems, the mechanical shaft to connect the steering column and steering pinion is removed. The steering input from the driver is no longer transmitted by the physical connection, but through the by-wire electronic communication. The advantages of SBW systems in road vehicles are to improve handling performance and control response, and to enhance the safety and comfort of drivers.

This book focuses on state-of-the-art drive-by-wire technology, particularly the control and evaluation technologies of the steer-by-wire system. Steer-by-wire systems replace the conventional mechanical linkages with electronic sensors, controllers, and actuators. Steer-by-wire systems offer the potential to enhance steering functionality by enabling features such as variable steer ratio, customizable road feeling and advanced vehicle dynamics control. Steer-by-wire systems will eventually replace the current traditional mechanical steering system, and has gradually been accepted by the automotive industry.

Based on the introduction of the history of X-by-wire technologies, what is steer-by-wire technology and why we need the chassis-by-wire vehicle are first discussed in this book. Compared to conventional mechanical steering systems, the main challenges of a steer-by-wire system are the tracking control and road feeling control in a fully decoupled structure. This book gives a complete explanation of the control technologies of the steer-by-wire system, from both forward dynamics and backward dynamics. Moreover, a shared steering control concept is introduced to improved vehicle handling performance.

The chief editor of this book is Prof. Xiaodong Wu, and the associate editor of this book is Dr. Liang Yan, both from Shanghai Jiao Tong University. The book was written by Chengrui Su, Chongyang Wang, Mingming Zhang, Shuhan Liu, Wenqi Li, etc., and edited by Prof. Xiaodong Wu. The content in this book involves the research results of doctoral and master's theses and has received strong support from graduate students in the research group. We would like to express our heartfelt thanks. Besides, many documents were cited in the writing process, and we would like to thank the authors of the relevant references again.

It is inevitable that there are some possible issues for further discussion in the content of this book. If you encounter any such issues, please feel free to contact us.

Brief Biography

Xiaodong Wu received the BS and ME degrees in mechanical and electronic engineering science from the China University of Petroleum, Beijing, in 2005 and 2008, respectively, and a PhD degree in mechatronic engineering from Ritsumeikan University, Kyoto, in 2011. He is currently a professor with the School of Mechanical Engineering, Shanghai Jiao Tong University. He is also member of Institute of Electrical and Electronics Engineers (IEEE) Vehicular Technology Society and IEEE Industrial Electronics Society. His current research interests include vehicle chassis control and the X-by-wire technology of intelligent vehicles. He has hosted more than 20 items of the National Key R&D Plan, the National Nature Science Foundation, and Provincial Research Projects. He has published over fifty papers as the first author and corresponding author; one monograph; and has been authorized over twenty invention patents.

1 Vehicle Steering System
An Introduction

1.1 OVERVIEW OF THE STEERING SYSTEM

Nowadays, with the improvement of people's living standards and the steady growth of the city, the car plays an increasingly important role in travel, and traffic has become more and more important to people. As a critical part of the vehicle, the steering system ensures that the movement of the vehicle is in line with the driver's lateral control intention and maintains the overall stability of the vehicle in case of deviation from the driving direction. Therefore, the steering system directly affects driving safety and the handling stability of the vehicle, playing a very important role in enhancing the driver's driving comfort.

Designing a good steering system is always the goal of automotive manufacturing. An excellent system can achieve light steering at low speed and stable steering at high speed – that is, at low speed, the driver can easily turn the steering wheel to make the car move at a wide angle, while at high speed, the driver will not experience any steering drifting. When designing the steering system, it is necessary to ensure that it can quickly and accurately adjust the wheel angle of the car according to the driver's intention, working safely and reliably, and with high stability. It is also necessary to ensure that the steering system provides the driver with a good sense of road under various conditions: normal steering, fast steering, wide-angle steering, and steering on slippery roads. In addition, the steering system should be easy to operate and have good maneuverability while guaranteeing light steering at low speed and stable steering at high speed.

As for the vehicle's motion control, with the development of computer science and control technology, various chassis-control systems have emerged. Their impact on the performance of vehicles and the interrelationship are shown in Figure 1.1.

The steering system directly affects the *handling stability* of the vehicle, which is different from the driving system, brake system, and suspension system. With the relentless development of the automobile, competition within the automotive industry has become increasingly fierce, and manufacturers have begun to invest vast resources in the research and development of new steering systems. Thanks to the development of microelectronics, artificial intelligence, and the penetration of

DOI: 10.1201/9781003481669-1

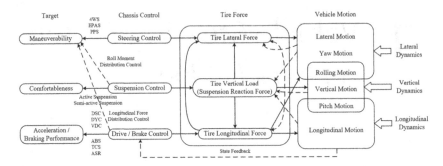

FIGURE 1.1   Chassis control systems in relation to vehicle dynamics.

Source: Yoshimi Furukawa & Masato Abe, *Advanced Chassis Control Systems*, 1997.

the Internet into the automotive field, new steering products are developing rapidly and have become essential subjects of researchers and scholars around the world.

1.2   HISTORY OF THE STEERING SYSTEM

The modern automotive steering system has been developing for more than a hundred years, since its birth. Its development history can be roughly divided into five stages, namely, mechanical steering, hydraulic power steering (HPS), electric hydraulic power steering (EHPS), electric power steering (EPS), and steering by wire (SBW). The vehicle steering system is developing in the direction of electronics and intelligence, and brings great convenience to people's driving experience, which makes driving ever more stable and safer.

In the 1880s, the initial vehicles used a mechanical steering system to realize the original function of changing the driving direction of the car. In the mechanical steering system, steering power is entirely provided by the driver to turn the steering wheel. If this type of vehicle is heavily loaded, the driver will feel heavy steering and is subject to driving fatigue. In addition, to improve the steering torque, the structure of mechanical steering system is very large, taking up a lot of space in the car, and reducing the driver's personal comfort.

With the development of automobile steering technology, since the 1950s hydraulic power steering and electric hydraulic power steering systems gradually replaced the traditional mechanical steering system. A hydraulic power steering system uses hydraulic pressure as auxiliary power, which reduces the torque of the driver turning the steering wheel and improves driving comfort. The electric hydraulic power steering system introduces electronic control and uses an electric motor to achieve different strengths of hydraulic power at different speeds.

Since the 1980s, the electric-power steering system has gradually developed and matured. An electronic control unit and electric motor are added to the traditional mechanical steering system, in which the electronic control unit controls the steering-actuating motor and applies appropriate torque to assist the driver to complete the steering operation. Compared with a hydraulic power steering system, the electric power steering system provides a more compact structure for small household vehicles, which had the advantages of a low failure rate and low cost.

In recent years, the emerging steer-by-wire technology has become a popular topic for researchers and scholars around the world. The steer-by-wire system mainly consists of the steering wheel unit, controller unit, and steering actuator unit. All these units also have a corresponding fault-tolerant control system, sensors, power supply, and other auxiliary devices. The steer-by-wire system eliminates the mechanical connection between the steering wheel and the steering actuator module, which uses electrical signals to achieve desired steering actuation. The significance of steer-by-wire is to meet the development requirements of intelligence and electrification in modern automobiles.

Based on the above discussion, the future steering system, as a part of automatic driving, will develop toward intelligence and strive for the goal of freeing people's hands. Next, we will introduce the structure, working principle, and the advantages and disadvantages of different types of steering systems to better explain the development of automotive steering systems.

1.2.1 Mechanical Steering System

All the force transmission components of the mechanical steering system are mechanical components, consisting mainly of a steering manipulator, steering gear, and steering transmission mechanism – specifically including the steering wheel, steering shaft, universal joint, steering knuckle, and steering cross tie rod, as shown in Figure 1.2.

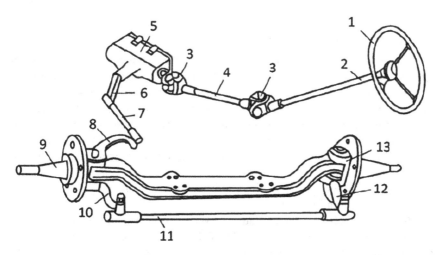

FIGURE 1.2 Structure of mechanical steering system.

Source: Zhang, Y. 2020; Technical analysis of wheeled vehicle steering system. *Journal of Physics.*

In the mechanical steering system, steering power is provided entirely by the driver without the help of external power. Mechanical steering gears can be divided into five main types: rack and pinion steering gears, recirculating ball steering gears, worm pin steering gears, worm roller steering gears, and worm gear steering gears. Among them, the most widely used systems are rack and pinion type and recirculating ball type.

The rack and pinion steering gear consists of the steering wheel, steering column, steering pinion, rack, tie rod, universal joint, etc. One end of the steering gear is connected to the steering column to receive the steering force input from the driver, and the other end engages with the steering rack to form a transmission pair, which drives the tie rod to turn the steering knuckle and adjust the wheel angle. The characteristics of rack and pinion steering gear are rigid, light, compact, and low cost. This kind of mechanism makes it easy to transfer the reaction force from the road to the steering handwheel, and it is good for the driver feeling the road. The rack and pinion steering gear is generally widely used in passenger cars.

The recirculating ball steering device mainly consists of a screw, nut, housing, and several small steel balls, and the screw is connected to the steering column. After the screw is rotated by the steering column, the nut is pushed up and down, driving the gears to make the steering arm reciprocate to complete the steering wheel's operation. The recirculating ball refers to the small steel balls between the nut and the screw, which can change the sliding friction between the nut and the screw into rolling friction with less resistance, so it has high transmission efficiency. Moreover, the recirculating ball steering gear has a strong load capacity and service life, so it is mostly used for heavy vehicles with large loads.

1.2.2 Hydraulic Power Steering System

In the previous subsection, we mentioned that the mechanical steering system realized the function of steering control, but the driver needs to overcome great steering resistance when the load of the vehicle is large. To solve this problem, the hydraulic power steering system (HPS) was born in the early twentieth century. The hydraulic power steering system has been in use for almost a century, and it is still widely used due to its reliable work and mature technology. The hydraulic power steering system is based on the mechanical steering system with the addition of a hydraulic system, including a hydraulic oil pump, hydraulic oil pipe, integral steering gear containing a hydraulic cylinder and hydraulic fluid control valve, and so forth, as shown in Figure 1.3.

The steering action of the hydraulic power steering system is still handled by the driver, but compared with a mechanical steering system, the power source acting on the steering mechanism is changed from completely human power to human power and hydraulic power together. HPS can reduce the torque required to turn the steering wheel and saves the driver's strength, improving driving comfort.

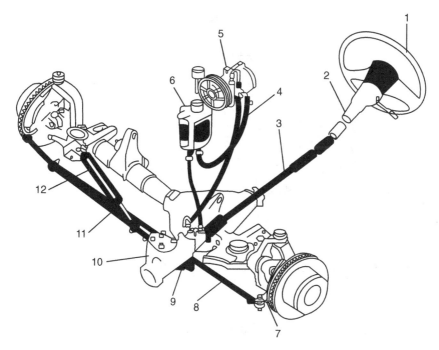

FIGURE 1.3 Structure of hydraulic power steering system.
Source: Figure created by authors.

When the engine is started, the hydraulic oil pump starts to work. The hydraulic oil pump is installed on the car engine to achieve the desired oil pressure. When the steering wheel is not turning, the hydraulic oil in the hydraulic pump flows directly back to the steering oil tank by the hydraulic fluid control valves. In this condition, the HPS does not produce the effect of power assistance. When the steering wheel turns to the left, the hydraulic fluid control valves open to connect the left oil pipe. The hydraulic oil flows to the left hydraulic cylinder, and HPS achieves power-assisted steering by pushing the piston and the steering screw to the right. It is the same when the steering wheel turns to the right.

1.2.3 ELECTRIC HYDRAULIC POWER STEERING SYSTEM

The electric hydraulic power steering system (EHPS) is developed from the traditional hydraulic power steering system. It is the traditional hydraulic power steering system with the addition of an electric motor, sensors, solenoid valves, and electronic control unit (ECU). The electric motor drives the hydraulic oil pump. The sensors detect the speed signal of the vehicle and the position signal of the steering wheel, then sends to the ECU as control inputs. The solenoid valves are operated by the ECU to achieve the desired output of the hydraulic oil pump.

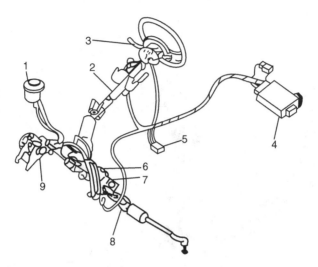

FIGURE 1.4   Structure of electric hydraulic power steering system.

Source: Figure created by authors.

The main structure of the electro-hydraulic power steering system is shown in the Figure 1.4.

Unlike the traditional hydraulic power steering system, which uses the engine to provide power, the electric hydraulic power steering system uses an electric motor to run the hydraulic oil pump. When the vehicle is being driven, the speed sensor detects the speed signal and transmits it to the ECU to identify the current driving status of the vehicle. The ECU controls the current in the coil of the solenoid valve to regulate the output of the hydraulic oil pump and, finally, acts on the steering actuator to realize the power-assisted steering.

The most distinctive feature of the EHPS is that the amount of assisting power is adjusted by the vehicle's velocity and steering speed. The driver load of the steering is also changed according to the velocity and steering speed. So the rotation speed of the motor is controlled to increase or decrease the output of the hydraulic oil pump and achieves a varying level of power assistance. If there is no steering control, the motor runs at a very low speed or stops working to reduce energy consumption.

1.2.4 Electric Power Steering System

If we say that EHPS is the initial application of the electrically controlled steering system, then electric power steering (EPS) is the mature application of the electronic control system in the vehicle's steering system. The electric power steering system does not use the hydraulic device of the hydraulic power steering system. The electric system is composed of sensors, electronic control unit (ECU), motors, and a reduction mechanism for the mechanical steering system. The sensors include

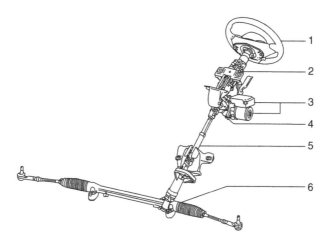

1
2
3
4
5
6

FIGURE 1.5 Structure of electric power steering system.
Source: Figure created by authors.

the speed sensor, torque sensor, pinion position sensor, and so forth, which can transmit vehicle driving parameters to the ECU as control decision parameters. The structure of an electric power steering system from Mazda Motor Corp. is shown in Figure 1.5.

When the vehicle is starting up, the driver rotates the steering wheel, and the torque sensor installed on the steering column detects the magnitude and direction of the torque applied by the driver. The signals from the torque sensor and the speed sensor are transmitted to the electronic control unit as control inputs. The ECU then analyzes and processes these sensor signals and uses certain control algorithms to determine the amount of torque required for power assistance. The output of ECU sends the torque request to the motor to generate the corresponding torque for the power steering assistance. When the vehicle speed increases, the ECU controls the motor to reduce the assisted torque and improve the steering stability at high driving speeds. If the vehicle's speed is too high, the EPS system exits the power-assisted mode and changes to manual steering. If the car is driving in a straight line without steering control, the motor stops working to save energy.

Depending on the motor installation position and mechanical structure, electric power steering can be generally divided into four categories: Column EPS (C-EPS), Pinion EPS (P-EPS), Dual-Pinion EPS (DP-EPS), and Rack EPS (R-EPS). The different types of electric power steering system are shown in Figure 1.6.

In C-EPS, the electric motor is mounted on the steering column, and the assisted torque is applied to the steering column. In P-EPS, the electric motor and reduction mechanism are mounted on the steering pinion, and the assisted torque is directly applied to the steering pinion. In DP-EPS, the electric motor and reduction

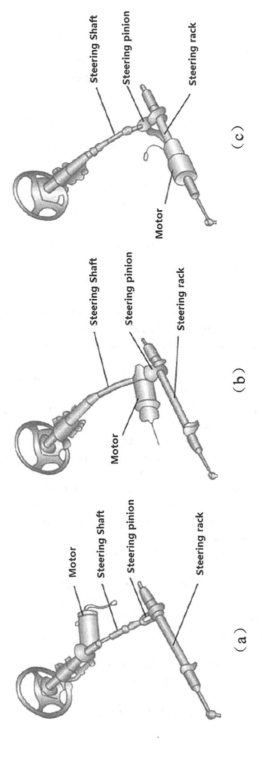

FIGURE 1.6 Different types of the electric power steering system.

Source: Zhao, W. et al., 2022. *Nonlinear Stability Control of Steer-by-Wire System.*

TABLE 1.1
Overview of Different Types of Electric Power Steering System

Items	C-EPS	P-EPS	DP-EPS	R-EPS
Location of motor	Steering column	Steering pinion	Additional pinion on the steering rack	Steering rack
Distance between motor and steering gear	Far	Medium	Near	Near
Cost	Low	Medium	High	Very high
Application models	Small-sized	Medium-sized	Upscale medium-sized	Large-sized, Performance-oriented
Advantages	• Simple structure • Easy to maintain • Low sealing requirements	• High transmission efficiency • Low motor noise • Low vibration	• Better transmission efficiency • Higher power provided • Less motor noise	• Highest transmission precision and transmission efficiency • Highest power provided
Disadvantages	• Closer to the steering wheel • Torque fluctuation and working noise	• Harsher working environment • High sealing requirements	• Harsh working environment • Higher cost	• Harshest working environment • Highest sealing requirements Highest cost

Source: Table created by authors.

mechanism are also mounted on the pinion of the steering gear, but arranged on the opposite side of the steering pinion to constitute a dual pinion assistance. In R-EPS, the electric motor and reduction mechanism are mounted on the steering rack, and the assisted torque is applied to the steering rack directly.

Based on the above introduction of EPS, they are applied to various types of vehicles. For example, C-EPS is generally used in small and medium-sized cars, while P-EPS and DP-EPS are generally used in medium-sized cars, and R-EPS is generally used in medium-sized and large-sized cars and in some performance-oriented cars. An overview of different types of the electric power steering system is shown in Table 1.1.

1.2.5   STEER-BY-WIRE SYSTEM

With the development of electronic control technology, *by-wire systems*, which are derived from aerospace technology, have begun replacing mechanical and hydraulic parts on vehicles. One of these advanced control technologies is the steer-by-wire (SBW) system. In the steer-by-wire system, unlike the traditional steering system, the mechanical shaft to connect the steering column and steering pinion is removed. The steering input from the driver is no longer transmitted by the physical connection, but through the by-wire electronic communication. The concept of vehicle steer-by-wire was first proposed in the 1950s.Due to the technical restrictions at that time, the application of steer-by-wire did not begin to make much progress until the late twentieth century. The world's first production car with the steer-by-wire technology came from Nissan Motor Company in 2013.

The structure of the steer-by-wire system is shown in Figure 1.7. Compared with electric power steering, the SBW system removes the traditional mechanism connection between the handwheel and steering pinion. The SBW system is made up of two separate modules: the handwheel module interacting with drivers, and the steering actuation module for generating vehicle steering responses. The steering command from the driver and the feedback torque from the tires are both transmitted by the bilateral signal wires. During the design of SBW, two main issues need to be resolved by the controller. The first factor is that the steering actuator should generate the desired turn motion to follow the input from the

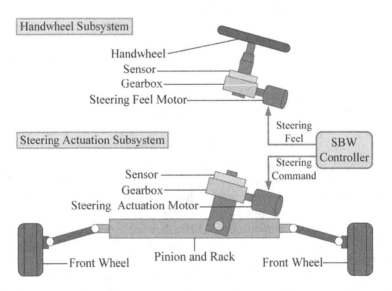

FIGURE 1.7   Structure of steer-by-wire system.

Source: Figure created by authors.

handwheel. The second is to simulate the road feel passed from the tire to steering rod via the handwheel module.

The advantages of SBW systems for road vehicles include improved handling performance and control response, enhanced safety and comfort of the drivers. In the SBW system, the physical linkage between the steering column and steering actuator is removed, so the steering command is no longer transmitted by the mechanism, but instead through the electronic signals by wire. By utilizing the characteristics of the SBW system, a customized steering dynamic can be designed with parameter modulation of the control system. For example. the steering control with a variable transmission ratio can be conducted through software, without any requirement of a hardware configuration.

However, since there is no mechanical connection between the handwheel and the steering pinion, if the electronic control module fails, the SBW system will lose effectiveness. A failure of the steering system will cause serious accidents. Therefore, the reliability and functional safety design of the SBW system are the key issues during the product development.

Based on the above introduction of various steering systems, the characteristics of these systems is summarized in Table 1.2.

1.3   DESIGN OF STEER-BY-WIRE SYSTEM

1.3.1   System Components

The composition of the steer-by-wire (SBW) system can be divided into mechanical and non-mechanical. The mechanical part includes the steering wheel module and the steering actuator module, while the non-mechanical part includes the control system, the fault-tolerant system, and the power supply system. The introduction of each component for the SBW system is described as follows.

1.3.1.1   Steering Wheel Module

The steering wheel module includes a steering handwheel, angle sensor, torque sensor, road-sensing motor, road-sensing motor reducer, and so forth. The module's functions are as follows: the sensor measures the steering wheel angle and steering wheel torque applied by the driver, and transmits these signals to the SBW controller to identify the driver's steering intention. At the same time, this controller receives the steering feeling torque signal from the SBW controller and drives the motor to generate the feedback torque on the steering handwheel. According to different application requirements, the steering wheel in the SBW system can also be changed to other manipulation mechanisms, such as joysticks, knobs, and so forth.

1.3.1.2   Steering Actuator Module

The steering actuation module mainly consists of the steering motor and its reducer, steering gear (steering rack and pinion), pinion angle sensors, and

TABLE 1.2
Comparison of Different Steering Systems

Categories	Advantages	Disadvantages
Mechanical Steering	• Simple structure • Low cost	• Demands driver's high steering torque, and leads to fatigue • Low steering accuracy on rough roads because of swinging vibration
HPS	• Reduces the driver's steering maneuvering force and improves driving comfort • Not easy to produce mechanical wear and tear, and no-lubrication, easy maintenance	• Difficulty in power-assistance regulation to adapt to different vehicle speeds • The maintenance of hydraulic pressure increases the fuel economy • The hydraulic system is noisy, and has high sealing requirements
EHPS	• Power performance can be adjusted according to the vehicle speed • The energy consumption can be improved by the electronic control • Make full use of traditional HPS system, has a good modular design	• Complex structure with high sealing requirements, risk of hydraulic oil leakage • Compared to the traditional HPS, the reliability of system is reduced
EPS	• Power performance can be adjusted according to the vehicle speed • Simple structure, save the interior space of the car • Dampens vibration and improves the driver's sense of road • Better fuel economy, no risk of hydraulic oil leakage, friendly to the environment	• The reliability of the electronic system under extreme working conditions is not as good as the traditional mechanical structure • The traditional 12V vehicle power supply system brings the bottleneck problem of motor power
SBW	• Improves driving safety, eliminates the frontal collision damage to the driver due to the removal of steering column • Improves the handling stability of the vehicle with variable ratio control • Customizable steering road feeling can be obtained to meet customers' individual needs	• Strict requirements for functional safety, and a more complex and rigorous fault-tolerant system is required. • High cost due to ECU, motor, and other redundant devices • The technology of road sense feedback is not mature

Source: Table created by authors.

front wheels. To make SBW have similar features as conventional steering systems, the control loop from handwheel to front wheel can be considered as a *teleoperation* application. The main issue is the tracking control accuracy of the steering actuator to follow the steering command from the steering wheel input. Moreover, redundant steering motors (such as the dual-winding motor) are employed on the steering actuator module, which improves the functional safety of steering control.

1.3.1.3 Control System

The SBW system is mainly composed of the SBW system controller, the SBW vehicle control unit (VCU), and the communication bus. The SBW system controller receives and processes the sensor signals to get the vehicle motion status, and sends motor control commands to the steering wheel module and steering actuator module. The communication bus is used for information interaction among the SBW system controller, the vehicle control unit, and other electronic modules. In addition, the SBW system controller also has to consider the fault-tolerant control due to the functional safety requirement.

1.3.1.4 Fault-Tolerant System

The fault-tolerant system is an important module of the SBW system, which includes a series of monitoring devices and fault processing algorithms. It can identify various fault types and fault levels, including the faults of system sensors, actuators, electronic control units, communication buses, and power supply systems. Fault-tolerant control can significantly enhance the safety and convenience of the driver and passengers and help prevent accidents. One may note that, in general, SBW system reliability enhancement using hardware redundancy leads to increased system volume, weight and cost; thus, this is not recognized as an optimized solution.

1.3.1.5 Power System

The power system is responsible for the SBW controller, two executive motors and other electrical modules. Generally, the maximum power of the front-wheel steering actuator is around 500 ~ 800W. And most of passenger cars have a low-voltage power supply of 12V and 3~3.5 kW. Because of the increased power load of the SBW system, it requires the support of a high-current discharge function. The performance of the power supply must be optimized to ensure the stable work of the SBW system under a heavy steering load.

1.3.2 SYSTEM LAYOUT

With the enrichment and continuous development of automotive SBW technology, a variety of typical steering actuator arrangements have evolved. In general, the existing wire-controlled steering systems can be divided into two categories: front-wheel SBW system and distributed SBW system.

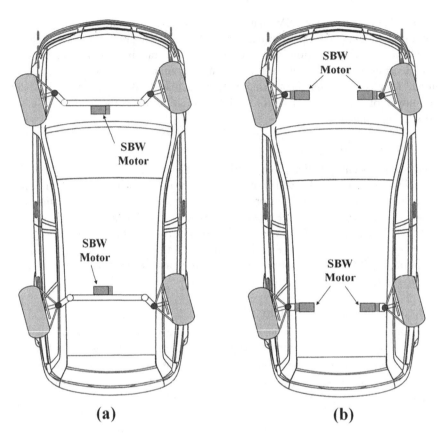

(a)   (b)

FIGURE 1.8 Two kinds of distributed SBW systems. (a) Independent steering for the front and rear axles. (b) Independent steering for all four wheels.

Source: Figure created by authors.

The front-wheel SBW system is developed from the traditional front-wheel mechanical steering system, and the main structure consists of two parts: the integrated part of the steering handwheel and steering column, and the integrated part of the steering actuator and front wheel. In accord with the different steering actuators, the front-wheel SBW system can be divided into the electric SBW system and the electro-hydraulic SBW system.

The distributed SBW system is divided into two different forms. The first form is independent steering for the front and rear axles, with the structural arrangement shown in Figure 1.8(a). The front and rear axles both have a steering actuator to turn the wheels. The second form is an independent steering system for all four wheels, with the structural arrangement shown in Figure 1.8(b). In this kind of structure, each wheel has a steering actuator to make independent turning control.

1.4 KEY TECHNOLOGIES OF THE STEER-BY-WIRE SYSTEM

Automotive SBW is a complex mechatronic system related to multiple technical areas. The most distinctive feature is that the steering operation of the driver and the SBW system are coordinated through the electronic control unit, without traditional mechanical connection. Therefore, the key technologies in the design of the SBW system include active steering angle tracking technology, road feeling control technology, and fault-tolerant control technology. The following gives an introduction to these key SBW technologies.

(1) Road feeling simulation-control technology

In order to achieve personalized steering feel of steer-by-wire vehicles, higher-fidelity steering feel models, which can create a wide variation of steering feel, are necessary. However, as model fidelity increases, the tuning of these models to obtain desired feel becomes more challenging. Therefore, finding the appropriate level of steering model fidelity is critical. The model must be complex enough to capture all the elements of steering feel that modern drivers care about, while the model must remain simple enough to be tuned intuitively [1].

In general, the steering feel model can be established through analyzing the dynamics of the steering system. By using vehicle information such as steering wheel angle, lateral acceleration, and steering motor current, the steering feel model can be built through the parameter fitting method, in which the steering motor torque is transferred directly to drivers by the steering motor current [2]. For example, Zheng et al. built a steering feel model by estimating the rack force through the Kalman filter [3]. Ma et al. built a steering feel model based on the information of the three-degree-of-freedom vehicle model and the magic tire model [4]. Wang et al. further introduced inertia, damping, and aligning torque to establish a steering feel model [5]. However, steering feels obtained by these methods are hard to adjust in accord with different driver styles. Artificial intelligence-based algorithms have been utilized in the feedback torque estimation in recent years. For instance, Van Ende et al. designed a feed-forward neural network (FNN) to obtain the steering feel feedback torque [6].

(2) Active steering angle tracking control technology

During the control of SBW system, the issue concerning this kind of decoupled system is the accurate tracking control of the steering actuator to follow the steering command from handwheel input. The traditional closed-loop PID controller [7]–[9] was introduced to solve this issue, but the integral term cannot avoid its inherent large overshoot and response delay. J.C. Gerdes [10] proposed a steering control model with acceleration, velocity, and friction feed-forward compensation to improve the SBW system response. A.E. Cetin in [11]–[12] improved this kind of model-based controller by combining it with an online adaptive parameter estimator. Since the external rack force in these researches are calculated with the dynamic model, it is difficult to avoid the influence of system disturbance in complex driving conditions. Hai Wang et al. proposed a sliding mode controller to deal with the model's uncertainties, which can enhance the algorithm robustness [13]–[15]. During the active steering angle-tracking control, the main challenge is the unpredictable system disturbances. Here the disturbances include the unknown external rack force, system friction and model uncertainties. Inspired by the concept of active suspension system in vehicles, if the rack force of SBW can be estimated and compensated, the robustness of steering track control can be improved.

The main advantage of the SBW system over the traditional mechanical steering system is that the dynamic transmission of the steering system can be completely decoupled. In the traditional steering system, the variable gear–ratio steering (VGS) control was considered as an advanced technology to improve vehicle cornering stability and handling performance, which has a changeable steering gain with respect to the different states of vehicle dynamics. Shimizu et al. found that the larger steering wheel angle manipulation could increase the driver's physical workload, and the meticulous steering operation could also increase the driver's mental workload [16]. A typical VGS system is an active front-steering system from BMW corporation, which utilizes a double planetary gear system and an electric actuator motor to facilitate driver independent steering of the front wheels [17]–[18]. This system is intended to improve steering stability at different driving speeds by an optimized transmission ratio. Because of the mechanical decoupled structure of SBW system, the steering control with a variable transmission ratio can be conveniently conducted through a software, without any requirement of a hardware configuration.

(3) Fault-tolerant control technology

Compared with a mechanical steering system, an electrical SBW system usually consists of a large number of electrical components that are more sensitive to control changes. Therefore, SBW systems have unpredictable failure characteristics, which may lead to serious results, including vehicle damage and passenger injury. Therefore, the SBW systems require a higher level of fault tolerance than do traditional steering systems [19].

To increase the reliability of SBW system, one common way is to offer improved reliability with full mechanical backup [20]. Infiniti Q50 deploys a number of backup systems, including a conventional mechanical steering linkage to protect the steering system against faults. A dual ECU-motors system is also a common method to improve reliability, and each motor has its own ECU module. If one motor fails, the corresponding ECU will shut down and the other motor control loop with the other ECU, will carry out all the steering-control operations. The advantages of hardware redundancy techniques are easy to install, and have high reliability and are simple to control. But, the mechanical backup systems are heavy and expensive and do not give enough freedom to use the potential of the electrical systems. In addition, the redundancy system – which consists of redundant actuators, ECU, power amplifier and other elements – brings high complexity in system design. To solve this issue, the steer-by-wire system can be designed and tested according to road vehicle functional safety standard ISO 26262.

The fault-tolerant control of the automotive SBW system mainly focuses on the diagnosis of power supply, sensors, actuators, electronic control units, and communication buses. The related technologies are composed of the fault detection, diagnosis and fault-tolerant control technology of the system. Among them, the fault detection technology should be able, in a timely way, to find the changes in the system, which is based on the sensory information of component status. The fault diagnosis technology should be able to determine the faulty component and the type of fault according to the above changes. The fault-tolerant control technology should be able to use the corresponding software or redundant hardware system to realize the function reconfiguration of the faulty component. The designed fault-tolerant control should have the ability to ensure basic steering requirements even under a low-level system failure.

REFERENCES

[1] Balachandran, A. and Gerdes, J. C. (2014). Designing steering feel for steer-by-wire vehicles using objective measures. *IEEE/ASME Trans. Mech.*, 20(1):373–383.

[2] Yang, S., Den, C., Ji, X. and Chen, K. (2007). *Research on road feeling control strategy of steer-by-wire*. SAE Paper No. 2007-01-3652.

[3] Zheng, H., Zong, C. and Yu, L. (2013). *Road feel feedback design for vehicle steer-by-wire via electric power steering*. SAE Paper No. 2013-01-2898.

[4] Ma, B., Liu, Y., Ji, X. and Yang, Y. (2018). Investigation of a steering defect and its compensation using a steeringtorque control strategy in an extreme driving situation. *Proc. Inst. Mech. Eng., Part D: J. Automob. Eng.*, 232(4): 534–546.

[5] Wang, J., Wang, H., Jiang, C., Cao, Z., Man, Z. and Chen, L. (2019). *Steering feel design for steer-by-wire system on electric vehicles*. Chinese Control Conf. (CCC), Guangzhou, China.

[6] Van Ende, K. T. R., Schaare, D., Kaste, J., Küçükay, F., Henze, R. and Kallmeyer, F. K. (2016). Practicability study on the suitability of artificial, neural networks for the approximation of unknown steering torques. *Veh. Syst. Dyn.*, 54(10):1362–1383.

[7] Oh, S.-W., Chae, H.-C., Yun, S.-C. and Han, C.-S. (2004). The design of a controller for the steer-by-wire system. *JSME Int. J. Series C Mech. Syst., Mach. Elements Manuf.*, 47(3):896–907.

[8] Amberkar, S., Bolourchi, F., Demerly, J. and Millsap, S. (2004). *A control system methodology for steer by wire systems*. SAE Technical Paper 2004-01-1106.

[9] Wang, X. et al. (2012). Bilateral control method of torque drive/angle feedback used for steer-by-wire system. *SAE Int. J. Passenger Cars Electron. Elect. Syst.*, 5: 479–485.

[10] Yih, P. and Gerdes, J. C. (2005, Nov). Modification of vehicle handling characteristics via steer-by-wire. *IEEE Trans. Control Syst. Technol.*, 13(6):965–976.

[11] Cetin, A. E., Adli, M. A., Barkana, D. E. and Kucuk, H. (2012). Adaptive online parameter identification of a steer-by-wire system, *Mechatronics*, 22(2):152–166.

[12] Cetin, A. E., Adli, M. A., Barkana, D. E. and Kucuk, H. (2010, Jan). Implementation and development of an adaptive steering-control system. *IEEE Trans. Veh. Technol.*, 59(1):75–83.

[13] Wang, H. et al. (2014, Nov). Robust control for steer-by-wire systems with partially known dynamics. *IEEE Trans. Ind. Inform.*, 10(4):2003–2015.

[14] Wang, H. et al. (2016, Sep). Design and implementation of adaptive terminal sliding mode control on a steer-by-wire equipped road vehicle. *IEEE Trans. Ind. Electron.*, 63(9):5774–5785.

[15] Wang, H. et al. (2014, March). Sliding mode control for steer-by-wire systems with ac motors in road vehicles, *IEEE Trans. Ind. Electron.*, 61(3):1596–1611.

[16] Shimizu, Y., Kawai, T. and Yuzuriha, J. (1999). Improvement in driver-vehicle system performance by varying steering gain with vehicle speed and steering angle: VGS (variable gear-ratio steering system). SAE technical paper 1999-01-0395.

[17] Serarslan, B., Bootz, A. and Schramm, D. (2010). Enhancement of steering and safety feeling in a steer-by-wire application. *IFAC Proc.*, 43:679–684.

[18] Kumar, E. A. and Kamble, D. N. (2012). An overview of active front steering system. *Int. J. Sci. Eng. Res.*, 6:1–10.

[19] Huang, C., Huang, H., Naghdy, F., et al. (2020). Actuator fault tolerant control for Steer-by-Wire systems. *Int. J. Control*, 94:3123-3134.

[20] Yao, Y., and Daugherty, B. (2007). *Control method of dual motor-based steer-by-wire system.* SAE Technical Paper.

2 Active Tracking Control for Steer-by-Wire System

2.1 BACKGROUND AND MOTIVATION

The steer-by-wire (SBW) system, functioning as the basic technology of the intelligent vehicle, is playing an important role in the development of modern vehicles. In the SBW system, compared with traditional mechanical steering, the physical linkage between the steering column and steering actuator is removed, so the steering command is no longer transmitted by the mechanism, but through electronic signals by wire. In order to make desired steering motions in a decoupling system, one of the most important problems is the angle tracking control of the SBW actuator.

Because of the nonlinear characteristic of vehicle dynamics, the core issue for the performance of the SBW system is the steering robustness against parameter variations, external disturbances, and road condition changes [1]. During the design of the controller, error-based tracking control of the steer-by-wire system is the main algorithm to realize the desired steering motion. Various control strategies have been applied to realize and improve the tracking control performance of the SBW system [2], [3], [4]. It can be found in previous studies that model-based methods are applied by most controllers. However, the model-based approach has a high requirement on the accuracy of system parameters. External disturbances and model nonlinearities can bring uncertainties of the system parameters. The measured system parameters that are assumed as constant input to the controlled model will reduce the robustness of tracking performance. How the SBW controller should be designed to achieve good steering performance is a challenging task, especially when the system parameters are unpredictably varying with uncertainties. Hence, an adaptive controller with online parameters estimation is necessary for the model-based approach.

2.2 MODELLING OF THE STEER-BY-WIRE SYSTEM

In order to have consistent features and dynamic characteristics, the architecture of SBW is defined in a mechanism similar to a traditional steering system. The SBW system is made up of two parts: the handwheel module and the steering actuator module. The handwheel module is shown in Figure 2.1, which includes a handwheel, a steering column, a sensory system, and a feedback motor. The

DOI: 10.1201/9781003481669-2

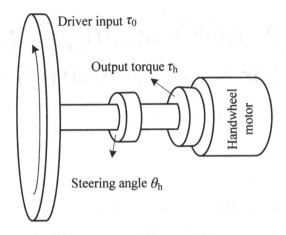

FIGURE 2.1   Handwheel module of steer-by-wire system.
Source: Figure created by authors.

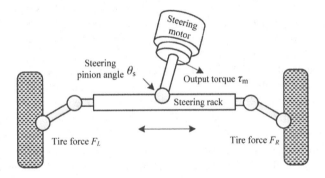

FIGURE 2.2   Steering actuator module of steer-by-wire system.
Source: Figure created by authors.

steering actuator module is shown in Figure 2.2, which consists of a steering actuator, steering rack and pinion, and a sensory system. In this chapter, the angular position of the handwheel is defined as the input signal, and the angular position of the automobile's front wheel is defined as the output signal of the system. To simplify the study, the steering pinion angle is defined as the output of the system and is equivalent to the angular position of the front wheel due to the fixed mechanical transmission ratio.

To build a dynamic model of the SBW system, it is assumed that the system is linear time-invariant and free of noise disturbance and un-modeled dynamics. The elements on the handwheel module and steering actuator module are represented by their total moments of inertia and viscous friction coefficients. To simplify the analysis of the steering system, we define the dynamic model of the SBW system with the handwheel module and steering actuator module. Based on the above analysis and assumptions, the handwheel module can be simplified as a one degree-of-freedom system, and its system dynamic model can be defined as a second-order differential equation according to Newton's laws. It can be given as

$$J_h \ddot{\theta}_h + B_h \dot{\theta}_h + \tau_{fh} \operatorname{sgn} \dot{\theta}_h = \tau_h + \tau_0 \qquad (2.1)$$

where J_h is the equivalent inertia of the handwheel module with respect to the column, B_h is the equivalent viscous friction, τ_{fh} is the coulomb friction, $\theta_h, \dot{\theta}_h, \ddot{\theta}_h$ is the angular position, velocity, and acceleration of the steering wheel respectively, τ_h is the controlled torque generated by the feedback motor, τ_0 is the input torque applied by the driver.

The system dynamic of steering actuator module can be described as a second-order differential equation:

$$J_s \ddot{\theta}_s + B_s \dot{\theta}_s + \tau_{fs} \operatorname{sgn} \dot{\theta}_s + (F_R + F_L)R_s = \tau_M \qquad (2.2)$$

where J_s is the equivalent inertia of the steering actuator module, B_s is the equivalent viscous friction, τ_{fs} is the coulomb friction, F_R and F_L are the feedback forces of the right and left tires, $\theta_s, \dot{\theta}_s, \ddot{\theta}_s$ is the angular position, velocity, acceleration of the steering pinion, τ_M is the controlled torque generated by the steering motor, R_s is the radius of pinion.

Considering the system dynamics in Eqns (2.1) and (2.2), the state space of the SBW system can be defined as

$$\begin{cases} \dot{x}_1 = x_2 \\ \dot{x}_2 = -P_{Bs} x_2 - P_{Fs} \operatorname{sgn} x_2 - P_{RL}(F_R + F_L) + P_M \tau_M \\ \dot{x}_3 = x_4 \\ \dot{x}_4 = -P_{Bh} x_4 - P_{Fh} \operatorname{sgn} x_2 + P_h(\tau_h + \tau_0) \end{cases} \qquad (2.3)$$

where the state variables x_1 and x_2 are the steering pinion angular position θ_s and velocity $\dot{\theta}_s$, x_3, and x_4 are the angular position θ_h and velocity $\dot{\theta}_h$ of the handwheel, the parameter P_{Bs}, P_{Fs}, P_{RL}, P_M, P_{Bh}, P_{Fh}, and P_h are B_s/J_s, τ_{fs}/J_s, $1/J_s R_s$, $1/J_s$, B_h/J_h, τ_{fh}/J_h, and $1/J_h$, respectively.

2.3   IDENTIFICATION OF SYSTEM PARAMETERS

To realize the desired steering motion on the SBW system, the model-based control approaches are employed by most previous studies. Based on the dynamic model of SBW in previous section, the mechanical parameters of SBW can be obtained by physical measurement. However, due to the existence of environment noise, model nonlinearities, electromechanical coupling and other disturbances, the system parameters are time-varying, which will decrease the performance of

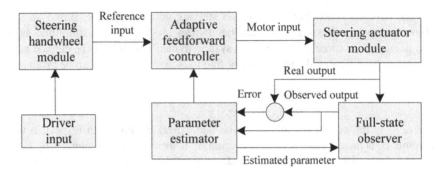

FIGURE 2.3 Architecture of the adaptive feedforward control with parameter estimator.
Source: Figure created by authors.

the model-based control algorithm. During model-based control, the identification of SBW system parameters directly influences the accuracy and efficiency of the steering performance. To solve this problem, an adaptive parameter identification method is introduced in this part to reduce the instability of control. Meanwhile, the parameter gradient projection method is applied to eliminate the parameter drift, while a full-state observer is used to reduce the influence of noise disturbance. The architecture of the adaptive feedforward control method is shown in Figure 2.3, which is a model-based approach with parameter identification [5].

2.3.1 Adaptive Online Parameter Identification

2.3.1.1 Output Error Identification

The output error identification method based on the state space of the SBW system, the following linear time-invariant single input, single output system is considered in Eqn. (2.4). The output error identification method is applicable to estimate the desired parameter P_a and P_b in the equation, and its estimating law is based on the state estimation error ε. According to the research in Ioannou and Sun [6], Sinha and Kuszta [7], by the input of excitation signal, the estimated parameters can converge to their expected values after iterative computations.

$$\dot{x} = -P_a x + P_b u$$
$$\varepsilon = x - \hat{x}$$
(2.4)

where x is the state variable, $\hat{x}$ is the estimated state variable, u is the input signal, P_a and P_b are unknown parameters.

Because tracking control is the main issue during the SBW control, so only the parameter identification procedure of the steering actuator module is considered in this part. Based on the state space of the SBW system in Eqn (2.3), the steering actuator module is presented as

$$\begin{cases} \dot{x}_1 = x_2 \\ \dot{x}_2 = -P_{Bs}x_2 - P_{Fs}\,\text{sgn}\,x_2 - P_{RL}(F_R + F_L) + P_M\tau_M \end{cases} \quad (2.5)$$

To take parameter identification of P_{Bs} and P_M firstly, a simplified model is designed without considering the tire forces. The parallel model configuration can be written as Eqn. (2.6), and the estimation error is given in Eqn. (2.7).

$$\dot{\hat{x}}_2 = -P_{Bs}\hat{x}_2 - P_{Fs}\,\text{sgn}\,\hat{x}_2 + P_M\tau_M \quad (2.6)$$

$$\varepsilon = x_2 - \hat{x}_2 \quad (2.7)$$

To get a stable parameter estimator, the adaptive law is selected as

$$\begin{aligned} \dot{\hat{P}}_{Bs} &= -\lambda_{Bs}\varepsilon\hat{x}_2, \quad \hat{P}_{Bs}(0) = \hat{P}_{Bs_0} \\ \dot{\hat{P}}_M &= -\lambda_M\varepsilon\tau_M, \quad \hat{P}_M(0) = \hat{P}_{M_0} \end{aligned} \quad (2.8)$$

where λ_{Bs}, λ_M are adaptive gains, which determines the response time and converge rate.

2.3.1.2 Parameter Gradient Projection

In real applications, the performance of the SBW system is affected by many factors, such as external noise disturbance, the nonlinearities of the model and so on. And these un-modeled factors will result in parameter drift, which drives the adaptive estimator to becoming unstable. To eliminate the parameter drift, a gradient projection method is introduced to keep the estimated parameter values within previously defined bounds in the parameter space. Based on the mechanical parameters supplied by the manufacturer, bounded uncertainties of each parameter were pre-defined in Eqn. (2.9). The adaptive estimator works when the estimated parameter value is within the pre-defined range. If the estimated value is beyond the bound, the adaptive estimator is set to zero and switched off to avoid the parameter drift.

$$\begin{aligned} \dot{\hat{P}}_{Bs} &= \begin{cases} -\lambda_{Bs}\varepsilon\hat{x}_2 & , \quad P_{Bs\_min} \le \hat{P}_{Bs} \le P_{Bs\_max} \\ 0 & , \quad \text{otherwise} \end{cases} \\ \dot{\hat{P}}_M &= \begin{cases} -\lambda_M\varepsilon\tau_M & , \quad P_{M\_min} \le \hat{P}_M \le P_{M\_max} \\ 0 & , \quad \text{otherwise} \end{cases} \\ \hat{P}_{Bs}(0) &= \hat{P}_{Bs_0}, \quad \hat{P}_M(0) = \hat{P}_{M_0} \end{aligned} \quad (2.9)$$

where P_{Bs_min}, P_{Bs_max}, P_{M_min}, and P_{M_max} are the boundaries of each interval.

2.3.1.3   Full-order State Observer

In order to have a stable and controllable estimator, a full-order state observer is introduced to improve the performance of the SBW control model. During the development of the state observer, a Luenberger observer is used to maintain the stability of the control system. The state equation of the proposed observer is given as

$$\dot{\hat{x}} = (A - GC)\hat{x} + Gy + Bu \tag{2.10}$$

Where the system matric A is $\begin{bmatrix} 0 & 1 \\ 0 & -P_{Bs} \end{bmatrix}$, control matric B is $[0 \ P_M]^T$, output matric C is $[1 \ 0]$, and the feedback gain matric G is $[g_1 \ g_2]^T$.

The parameter estimator serves as a subsystem of the state observer. The stability of the state observer can be ensured when the eigenvalues of a system are distributed at any location of complex plane. Since the estimated parameters are bounded, the key to designing a state observer is to determine the gain coefficient g_1 and g_2. The desired eigenvalues should have a large negative real part, which ensures the state observer system has a fast converge rate.

2.3.2   Adaptive Feed-Forward Control Method

To realize the tracking control of steering actuator module, a proportional derivative (PD) controller is designed by using the feedback of the steering angle and velocity. The motor torque can be expressed as

$$\tau_{PD} = K_p(\theta_r - \theta_s) + K_D(\dot{\theta}_r - \dot{\theta}_s) \tag{2.11}$$

where θ_r is the reference steering angle, K_p and K_D are the proportional and differential gains. Based on the proposed online parameter identification, the adaptive feed-forward control approach is combining with the PD control of the SBW system. According to Eqn. (2.2), the control law of adaptive feedforward is given as

$$\tau_M = \frac{1}{P_M}\ddot{\theta}_s + \frac{P_{Bs}}{P_M}\dot{\theta}_s + \frac{P_{Fs}}{P_M}\text{sgn}\,\dot{\theta}_s + \tau_{PD} \tag{2.12}$$

From Eqn. (2.12), P_{Bs}, P_M, and P_{Fs} are the parameters with bounded uncertainties. The online parameter identification is applied to identify the value according to the realistic SBW system. The adaptive estimator is conducted in real time to reduce parameter drift and external disturbance.

Based on the characteristics of the servo motor, the controlled output torque is proportional to the input current, which has $\tau_M = k_{SM}V_{SM}$. If we define output torque $\tau_M = k_{SM} V_{SM}$, and V_{SM} is the input current of the servo motor, the adaptive feedforward control law is rewritten as

$$V_{SM} = \frac{1}{P_{SM}} \ddot{\theta}_s + \frac{P_{Bs}}{P_{SM}} \dot{\theta}_s + \frac{P_{Fs}}{P_{SM}} \operatorname{sgn} \dot{\theta}_s + \frac{\tau_{PD}}{k_{SM}} \qquad (2.13)$$

where $P_{SM} = K_{SM} P_M$.

According to Eqns (2.5) and (2.6), the state equation of the system and the parallel model configuration can be modified respectively as

$$\begin{cases} \dot{x}_1 = x_2 \\ \dot{x}_2 = -P_{Bs} x_2 - P_{Fs} \operatorname{sgn} x_2 + P_{SM} V_{SM} \\ \hat{P}_2 = -\hat{P}_{Bs} \hat{x}_2 + \hat{P}_{SM} V_{SM}, \quad \hat{x}_2(0) = \hat{x}_{2_0} \end{cases} \qquad (2.14)$$

and the updated adaptive laws for the final controller can be expressed as

$$\dot{\hat{P}}_{Bs} = \begin{cases} -\lambda_{Bs} \varepsilon \hat{x}_2 &, \quad P_{Bs\_min} \leq \hat{P}_{Bs} \leq P_{Bs\_max} \\ 0 &, \quad \text{otherwise} \end{cases}$$

$$\dot{\hat{P}}_{SM} = \begin{cases} -\lambda_{SM} \varepsilon V_{SM} &, \quad P_{SM\_min} \leq \hat{P}_{SM} \leq P_{SM\_max} \\ 0 &, \quad \text{otherwise} \end{cases} \qquad (2.15)$$

$$\hat{P}_{Bs}(0) = \hat{P}_{Bs_0}, \quad \hat{P}_{SM}(0) = \hat{P}_{SM_0}$$

2.3.3 EXPERIMENTS AND ANALYSIS

To verify the proposed tracking control method, an experimental platform of SBW is shown in Figure 2.4. The control unit of hardware-in-the-loop (HIL) is a rapid-control prototype, which can build the control program by Matlab/Simulink models. Furthermore, to simulate aligning and resisting moments generated during steering motion, two coiled springs are installed on the both sides of the steering rack. The main specifications of the SBW experimental platform are listed in Table 2.1.

2.3.3.1 Parameter Identification of Target SBW System

Based on the above SBW experimental platform, the parameter identification for the steering actuator module is investigated. By using the mechanical parameters supplied by the manufacturer, the initial bounded uncertainties of $\hat{P}_{Bs}$ and $\hat{P}_{SM}$ is set as [2.0, 5.0] and [0.2, 2.0] respectively. Meanwhile, during the design of the state observer, the initial value P_{Bs} can be any value within the pre-defined boundary. To acquire faster convergence rate, the eigenvalues of the system matric A-GC of the state observer in Eqn. (2.10) should have large negative real parts. Based on the boundaries of PBs, the feedback gain g_1 and g_2 should be designed to ensure that

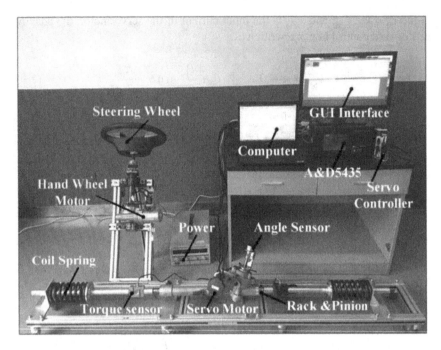

FIGURE 2.4   SBW tracking control experimental platform.

Source: Figure created by authors.

TABLE 2.1
Specification of the SBW Experimental Platform

Handwheel motor rated power:	600 w
Handwheel motor reduction ratio:	16.5
Steering servo motor rated power:	960 w
Steering servo motor reduction ratio:	21
Radius of steering pinion:	0.8 cm
Input DC voltage:	12 V
Angle accuracy of sensor:	< 1°
Range of force sensor:	± 1000 N
Accuracy of force sensor:	± 0.05 %F.S.

Source: Table created by authors.

the eigenvalues are located in the left part of the complex plane. The final design of g_1 and g_2 is 150 and 3000 respectively, which ensures the stability of the state observer.

For the experiments of the steering actuator module, the excitation signal of motor current is chosen as a sinusoidal signal $V_{sm} = 100V_0 \sin(\omega t)$, and the sampling interval is set as 1ms. During the parameter identification, several experiments have been done to study the performance of the adaptive estimator with different adaptive gains λ_{Bs}, λ_{SM}, and changed amplitude V_0, frequency ω of the excitation signals.

(1) Parameter identification with different adaptive gains

In this part, the influences of different adaptive gains during the parameter identification is studied by experiments. The excitation signal V_{sm} is defined as $V_{sm} = 1.5 \sin(\omega t)$, and the initial values of uncertain parameters are set as $\hat{P}_{Bs_0} = 3.5$, $\hat{P}_{SM_0} = 1$.

To check the convergence rate and stability of the estimator, three sets of adaptive gains are chosen. Here, λ_{Bs} is chosen as 500, 750, and 1000 respectively, and λ_{SM} is chosen as 250, 500, and 1000 respectively. The experiment results is shown in Figures 2.5 and 2.6. It can be found that, with the adaptive gains increase, the convergence rate of the estimator becomes faster, but also results in worse vibration. This is because the high gains speed up the response rate, but also result in instability of the adaptive estimator, which should be avoid in real applications.

(2) Parameter identification with different excitation signals

The results of estimated parameters P_{Bs} and P_{SM} with different excitation signals are investigated in this part. The adaptive gains are set as $\lambda_{Bs} = 750$, $\lambda_{SM} = 500$, and the initial value of the estimated parameters are $\hat{P}_{Bs_0} = 3.5$, $\hat{P}_{SM_0} = 1$. Figures 2.7

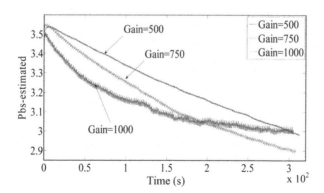

FIGURE 2.5 Estimated parameter $\hat{P}_{Bs}$ with different adaptive gains.
Source: Figure created by authors.

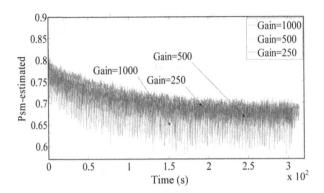

FIGURE 2.6   Estimated parameter $\hat{P}_{SM}$ with different adaptive gains.
Source: Figure created by authors.

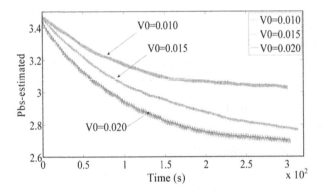

FIGURE 2.7   Estimated parameter $\hat{P}_{Bs}$ with different amplitude gain V_0.
Source: Figure created by authors.

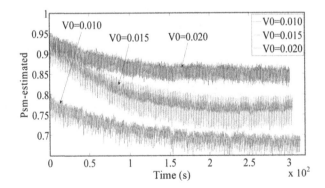

FIGURE 2.8   Estimated parameter $\hat{P}_{SM}$ with different amplitude gain V_0.
Source: Figure created by authors.

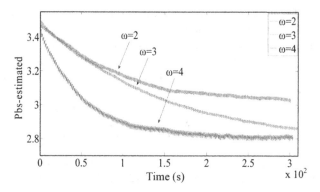

FIGURE 2.9　Estimated parameter $\hat{P}_{Bs}$ with different frequency ω (rad/s).
Source: Figure created by authors.

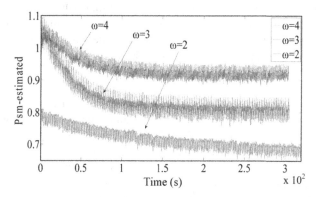

FIGURE 2.10　Estimated parameter $\hat{P}_{SM}$ with different frequency ω (rad/s).
Source: Figure created by authors.

and 2.8 show the results of changed amplitude V_0 from 0.01 to 0.02, while other variables are set constant. Similarly, Figures 2.9 and 2.10 show the results of changed frequency ω, which is chosen as 2, 3, and 4, respectively. It can be found that the increase of amplitude or frequency for excitation signals will obtain a smaller convergence value of $\hat{P}_{Bs}$ and a larger convergence value of $\hat{P}_{SM}$.

From the above experiment results can be drawn a conclusion that the system parameter P_{Bs} and P_{SM} have bounded uncertainties, which is varied with different adaptive gains and excitation signals. The convergence value of $\hat{P}_{Bs}$ varies from 2.7 to 3.2, $\hat{P}_{SM}$ varies from 0.6 to 1.0, both fluctuate in a smaller range than the initial defined bounded uncertainties.

The friction in the SBW system is generated from the relative movement of mechanical parts. To obtain the static friction by the experiment, it was observed that the level of motor current $V_{sm} = 0.2$ initiates the movement of steering pinion and rack without driver interaction. Hence, this is approximately the level of the servo motor torque that compensates the static friction and then initiates the steering motion. At the switching point of the steering actuator module from stop to move, the velocity of system can set to be zero. Thus, the dynamic model in Eqn. (2.14) is rewritten as

$$0 = -\hat{P}_{Fs} \operatorname{sgn} x_4 + \hat{P}_{SM} V_{SM} \qquad (2.16)$$

where $\hat{P}_{Fs} = \hat{P}_{SM} V_{SM}$. Because $\hat{P}_{SM}$ varies from 0.6 to 1.0, so upper and lower bounds for parameter $\hat{P}_{Fs}$ are designated as 0.12 and 0.20, respectively.

2.3.3.2 Tracking Control with Parameter Identification

From the results of the parameter identification, the upper and lower bounds of $\hat{P}_{Bs}$, $\hat{P}_{SM}$ and $\hat{P}_{Fs}$ are achieved in Table 2.2. By combining the feed-forward controller with adaptive online parameter estimation, the initial values for the estimated parameters are set within the defined bounds, shown in Table 2.2. The system parameters make online adjustment in terms of the identification results.

The experiments are conducted with simple proportional-derivative (PD) control and proposed adaptive feedforward (AF) control. The parameters and gains during the experiments are listed in Table 2.3. The tracking performance with different steering frequency are shown in Figures 2.11 and 2.13. Compared with PD controller, the tracking performance is improved significantly by proposed adaptive feedforward control. From the RMS error, the adaptive feedforward method has a reduced tracking error in high frequency. The tiny oscillation of the error curve is generated from the backlash of the gear system.

By the online estimation, the self-adaptive parameters are shown in Figures 2.12 and 2.14. The results show the robustness of the controller with the identification of unknown parameters during the dynamic steering motion. This method uses the estimated state variable instead of measured signal to the control model, which makes it less sensitive to the external disturbance. Moreover, it is easy to adapt to the real time application of the steering system, because it is based on system

TABLE 2.2
Bound Uncertainties of Estimated Values

Parameters	Upper and lower bounds	Initial values
$\hat{P}_{Bs}$	[2.7, 3.2]	3.0
$\hat{P}_{SM}$	[0.6, 1.0]	0.75
$\hat{P}_{Fs}$	[0.12, 0.20]	0.18

Source: Table created by authors.

TABLE 2.3
Parameters and Gains for SBW Experiments

Parameters	Upper and lower bounds
λ_{Bs}	1000
λ_{SM}	500
V_0	0.05
K_P	20
K_D	5

Source: Table created by authors.

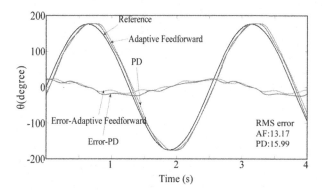

FIGURE 2.11 Tracking performance with low frequency $\omega = 2.5$ rad/s.
Source: Figure created by authors.

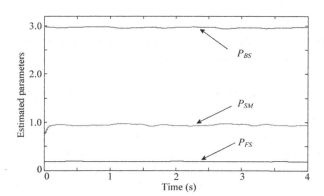

FIGURE 2.12 Estimated parameters with frequency $\omega = 2.5$ rad/s.
Source: Figure created by authors.

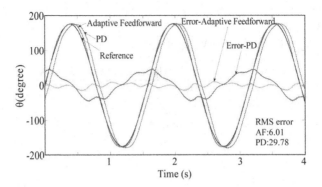

FIGURE 2.13   Tracking performance with high frequency ω = 4 rad/s.
Source: Figure created by authors.

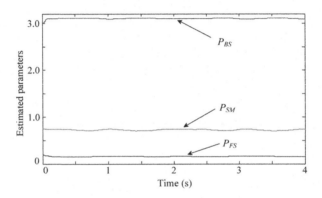

FIGURE 2.14   Estimated parameters with frequency ω = 4 rad/s.
Source: Figure created by authors.

state equation itself to calculate state estimation error, and needs less iterative calculation time during the parameter estimation.

2.4 ACTIVE TRACKING CONTROL FOR STEER-BY-WIRE SYSTEM

The architecture of the steer-by-wire system is shown in Figure 2.15. Compared with the traditional electronic power steering system or hydraulic power steering system, mechanical linkage between the steering column and steering actuator has been removed. Thus, the SBW system is made up of two separate modules: the handwheel module interacting with drivers and the steering actuation module for generating vehicle steering responses. The steering command from the driver and the feedback torque from the tires are both transmitted by the bilateral signal wires. To have an accurate steering performance, the tracking control of steering actuator system to follow the handwheel input is the key issue.

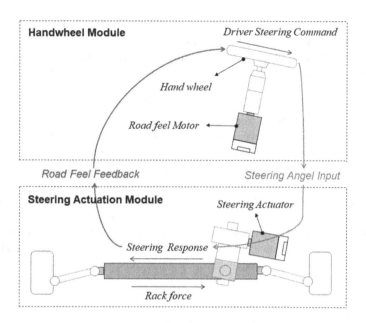

FIGURE 2.15 Architecture of steer-by-wire system.

Source: Figure created by authors.

At the same time, the steering tracking control algorithm should have sufficient stability margin and robustness to meet the vehicle's functional safety requirements. The failure forms of the actuator system in SBW are specifically manifested as understeering and oversteering. Understeering is generally caused by an insufficient response of the cornering amplitude or the serious lag of the cornering phase. The oversteering is typically due to the overshooting of the cornering amplitude. Both understeering and oversteering can cause control difficulty and become a psychological burden to the driver, and must be strictly averted. However, there is a great deal of randomness on the road. The material of the road surface, the degree of bumps, and the distribution of potholes – these are all unpredictable. The random road conditions interact with the tires, and the self-aligning torque of the tires passes through the steering rod and is finally loaded on the SBW system in the form of external rack force. From a control point of view, this rack force can be considered as a random and unknown disturbance. In an analogy to the traditional electric power steering (EPS) system, this disturbance is the main item of the external disturbance input to the steering execution system. Its magnitude is equivalent to the assistant torque of the motor. This disturbance not only seriously affects the kinematics of the system, but also loads the system in a random manner, which brings difficulties to the design of the steering-following controller of the SBW system.

Therefore, in order to improve the control accuracy, the steering-following controller is expected to have the ability to suppress parameter uncertainty, the model uncertainty, and also external disturbance. To solve the problem, we here introduce the concept of generalized rack force (GRF), which is defined as the sum of the rack force generated by the self-aligning moment, the parameter uncertainty and the model uncertainty. If the GRF can be estimated online and compensated for in real time, the steering-following performance of the system will be improved significantly.

In this chapter, to estimate the GRF we first design disturbance observer (DO) with three different structures, namely, linear disturbance observer (LDO), nonlinear disturbance observer (NDO) and extended disturbance observer (EDO). We will compare their estimated performance on the external disturbance and their engineering practicality. Then, based on the parameter identification algorithm, the adaptive pole configuration algorithm is designed, which aims to improve the dynamic response characteristics of the system by improving the accuracy of the model parameters, thereby improving the tracking accuracy of the steering system. Finally, we perform feed-forward compensation on the robust sliding mode controller with the estimated results of the DO, and propose an active tracker adapted to the SBW system. By strengthening the interference suppression ability of the algorithm, the steering-following performance of the system is further improved.

2.4.1 Design of Disturbance Observer

LDO is a Luenberger-like observer based on the state space. Its feedback gain matrix is a linear constant matrix. The convergence rate of the observer can be designed by configuring the poles of its system matrix. The structure of LDO is simple, and the parameter setting is convenient. NDO takes the nonlinearity of external disturbance into consideration and introduces an optimal nonlinear feedback function to achieve good estimation of the external disturbance. EDO integrates the advantages of the above two. By designing a suitable nonlinear gain, the estimation performance of EDO can be further improved. Meanwhile, it also has great engineering practicability.

As for designing the DO and steering-following controller, it is necessary to consider the influence of external rack force input and model uncertainty. According to previous experimental results, the equivalent moment of inertia J and the viscous friction coefficient B of the system are bounded. Assume that the model parameters of the system satisfy the following boundary constraints

$$\left| J - J_0 \right| \leq \Delta J$$
$$\left| B - B_0 \right| \leq \Delta B \tag{2.17}$$

where J_0 and B_0 are the nominal value of system parameters. The system parameters of the SBW plant model and the parameters of disturbance observers are given in Table 2.4 and Table 2.5.

With the convenience of DO design, we define $d(t)$, as the sum of the coulomb static friction item, the model uncertainty item and the rack force in the steering

TABLE 2.4
Parameters of SBW System

Symbol	Value	Unit
J	0.14	kg·ms^2
B	0.8	N·s
τ_f	3	N·m
i_{mc}	26.15	
i_{rc}	8e-3	m/rad

Source: Table created by authors.

TABLE 2.5
Parameters Design of Simulation Experiments

Parameters	Value
P1	5.71
P2	186.79
P3	7.14
G	$\begin{bmatrix} 100 & -280 & 40 \\ 1 & 200 & -160000 \end{bmatrix}^T$
K	[5 10]
$\begin{bmatrix} \kappa_1 & \kappa_2 & \kappa_3 & \delta \end{bmatrix}$	[6 11 6 0.01]
Kp	20
Ki	0.01
Kd	5
Ks	100
Λ	10
Tf	0.01

Source: Table created by authors.

dynamic. If we assume that $d(t)$ is a smooth variable and its first-order differential is bounded, then the dynamic equation of the system can be written concisely as

$$\ddot{\theta} = -p_1\dot{\theta} + p_2\tau_m - p_3 d \tag{2.18}$$

where $p_1 = B/J$, $p_2 = i_{mc}/J$, p3 = 1/J, and p1, p2, p3 are constrained by following bounded conditions

$$|p_1 - p_{10}| < \Delta p_1$$
$$|p_2 - p_{20}| < \Delta p_2 \qquad (2.19)$$
$$|p_3 - p_{30}| < \Delta p_3$$

Notice that the dynamic model in Eqn. (2.18) is more consistent with the actual working conditions of SBW system, because it considers the input of external rack force. Now, the state space expression of SBW system can be written as

$$\begin{cases} \dot{x} = Ax + B\tau_m + Ed \\ y = Cx \end{cases} \qquad (2.20)$$

where $x = \begin{bmatrix} x_1 \\ x_2 \end{bmatrix}$, $A = \begin{bmatrix} 0 & 1 \\ 0 & -p_1 \end{bmatrix}$, $B = \begin{bmatrix} 0 \\ p_2 \end{bmatrix}$, $E = \begin{bmatrix} 0 \\ -p_3 \end{bmatrix}$, $C = \begin{bmatrix} 1 & 0 \end{bmatrix}$.

2.4.1.1   Linear Disturbance Observer

LDO is a linear observer designed based on the system state space. Thanks to its simple structure, clear parameter setting criteria and good practicability in engineering design and applications, LDO has been developed as a mature technology for rack force estimation in the industry. Here it was first applied to the rack force estimation of the steering system.

Specifically, derived from the full-order Luenberger state observer, LDO considers GRF of the system $d(t)$ as external disturbance, denoted by the state variable xd, and adds it into the original state vector x, leading to a new state vector $[x_1, x_2, x_d]^T = [x, x_d]^T$. Assuming that d(t) slowly varies with respect to the controller dynamics, the new state variable xd will satisfy the following autonomous system

$$\begin{cases} \dot{x}_d = A_d x_d \\ \dot{y}_d = C_d x_d \end{cases} \qquad (2.21)$$

where $A_d = 0$, $C_d = 1$.

Combining the autonomous system (2.21) representing the dynamic characteristics of x_d with the original state space system (2.20) results in

$$\begin{cases} \begin{bmatrix} \dot{x} \\ \dot{x}_d \end{bmatrix} = \underbrace{\begin{bmatrix} A & EC_d \\ 0 & A_d \end{bmatrix}}_{A_l} \begin{bmatrix} x \\ x_d \end{bmatrix} + \underbrace{\begin{bmatrix} B \\ 0 \end{bmatrix}}_{B_l} \tau_m \\ y = \underbrace{\begin{bmatrix} C & 0 \end{bmatrix}}_{C_l} x \end{cases} \qquad (2.22)$$

Then the dynamic system (2.22) can be expressed by system matrix A_l, input matrix B_l and output matrix C_l, denoted by R(LDO) = [A_l, B_l, C_l].

Before designing the DO, that we need to prove the system (2.22) is observable. Construct the system observability matrix N_1, then

$$rank\,N_\ell = rank \begin{pmatrix} C_\ell \\ C_\ell A_\ell \\ C_\ell A_\ell^2 \end{pmatrix} = rank \begin{pmatrix} 1 & 0 & 0 \\ 0 & 1 & 0 \\ 0 & -p_1 & -p_3 \end{pmatrix} \qquad (2.23)$$

Since p1 and p3 are not equal to zero, Nl has full rank, indicating that the system is completely observable. Then R(LDO) can be regarded as a normal linear system, and the full-dimensional state observer can be derived. In order to distinguish the dynamics of the original system from that of external disturbances in the observer, we construct the LDO as a block matrix expression as shown in Eqn. (2.24). The schematic diagram of the LDO is shown in Figure 2.16.

$$\begin{cases} \begin{bmatrix} \dot{\hat{x}} \\ \dot{\hat{x}}_d \end{bmatrix} = \underbrace{\begin{bmatrix} A - G_1 C & EC_d \\ -G_2 C & A_d \end{bmatrix}}_{A_{LDO}} \begin{bmatrix} \hat{x} \\ \hat{x}_d \end{bmatrix} + \begin{bmatrix} B \\ 0 \end{bmatrix} \tau_m + \begin{bmatrix} G_1 \\ G_2 \end{bmatrix} y \\ \hat{y} = \begin{bmatrix} C & 0 \end{bmatrix} \begin{bmatrix} \hat{x} \\ \hat{x}_d \end{bmatrix} \end{cases}$$

$$(2.24)$$

where $G = [G_1, G_2]_T$ is the feedback gain matrix to be designed, ALDO is the system matrix of LDO.

Now let's discuss the stability and convergence of the LDO. It can be seen from Eqn. (2.24) that the last column of the system matrix ALDO is always

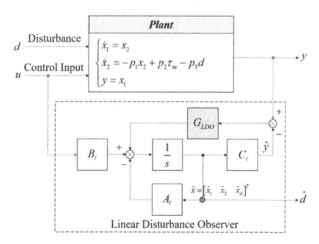

FIGURE 2.16 Structure of Linear Disturbance Observer.

Source: Figure created by authors.

non-zero. Meanwhile, since the first two columns of ALDO are related to G_1 and G_2, an appropriate feedback gain matrix G can always be worked out to make the first two columns of ALDO linearly independent, and finally make ALDO full-rank. Furthermore, an appropriate G can always be designed so that all of the eigenvalues of ALDO are distributed on the left side of the complex plane, ensuring the stability of the LDO. At this time, the estimated value of LDO exponentially converges with the true value. The convergence rate is positively correlated with the distances between the eigenvalues and the imaginary axis. The negative real part of eigenvalues can be configured through G.

It can be found that the structure of LDO is simple and linear. The excitation inputs of the observer are the torque input τ_m and the angle signal x_1 which are easily obtained. Its estimation performance or convergence speed can be configured and adjusted by designing the feedback gain matrix G. Besides, no other complicated calculation or analysis is required, so LDO is functional in engineering practice. However, the disadvantages of LDO are also obvious. Note that $A_d = 0$, then the trace of ALDO is equal to the trace of A-G1C. That is to say, the sum of the magnitudes of all negative real parts of ALDO is completely determined by that of A-G_1C. In this case, if we want to balance the convergence rate of each state variable in the LDO system, some elements in G will be set as very large numbers, which will bring a burden to the calculation and storage of the controller, and also amplify the noise signal in the system and affect the estimation performance of the observer. In addition, G will not be updated once it is set. A fixed gain means fixed convergence rate, so it does not have the optimal estimation ability under all conditions, especially for extreme working conditions.

2.4.1.2   Nonlinear Disturbance Observer

The core of NDO is to estimate the external disturbance by constructing a nonlinear mathematical observation structure. The concept was first proposed by Wenhua Chen et al., who used it to solve the disturbance suppression problem in the fields of aircraft attitude control and robot control, and achieved good results in engineering practice. Inspired by that, here we design an NDO for the SBW actuation system, to estimate the external rack force. We will also give the stability proof and convergence analysis of the algorithm then.

Here we rewrite Eqn. (2.18) as follows

$$d = \frac{-\ddot{x}_1 - p_1\dot{x}_1 + p_2\tau_m}{p_3}$$
$$= \frac{-\dot{x}_2 - p_1 x_2 + p_2\tau_m}{p_3} \tag{2.25}$$

Theoretically, the feedback gain function can be designed as $l(x_1, x_2)$ and the observer has a form like

$$\hat{d} = l\left(x_1, x_2\right)\left(d - \hat{d}\right)$$

$$= -l\left(x_1, x_2\right)\hat{d} + l\left(x_1, x_2\right)\left(\frac{-\dot{x}_2 - p_1 x_2 + p_2 \tau_m}{p_3}\right) \qquad (2.26)$$

$l(x_1, x_2)$ is used to eliminate the error of the estimated value. For the first-order differential equation as shown in Eqn. (2.26), an appropriate feedback $l(x_1, x_2)$ can always be designed so that the estimated value converges to the true value. However, the above-mentioned observer cannot be applied directly to the SBW system, because the steering hardware is usually equipped with a single angle sensor. Moreover, the chassis environment of the steering system is harsh and full of noise, so the acceleration signal required in the equation cannot be measured directly. Although the observer in Eqn. (2.26) cannot be used, it provides a theoretical basis for the design of the NDO in the following text.

To solve the problem of obtaining the angular acceleration signal, we define an auxiliary variable q(t) as

$$q(t) = \hat{d}(t) - g\left(x_1, x_2\right) \qquad (2.27)$$

where $g(x_1, x_2)$ is a nonlinear function to be designed.

Let the feedback function $l(x_1, x_2)$ satisfy Eqn. (2.28). Although matrix operations are involved, since both sides are scalars, the equation must hold.

$$\underbrace{l\left(x_1, x_2\right)\dot{x}_2}_{Scalar} = -p_3 \frac{\partial g\left(x_1, x_2\right)}{\partial x}\dot{x} = -p_3 \underbrace{\left[\frac{\partial g\left(x_1, x_2\right)}{\partial x_1} \quad \frac{\partial g\left(x_1, x_2\right)}{\partial x_2}\right]\begin{bmatrix} x_2 \\ \dot{x}_2 \end{bmatrix}}_{Scalar} \qquad (2.28)$$

The dynamic equation of q(t) can be obtained by considering the above equations, as shown below

$$q = \hat{d} - \frac{d}{dt}g\left(x_1, x_2\right)$$

$$= \hat{d} - \left[\frac{\partial g\left(x_1, x_2\right)}{\partial x_1} \quad \frac{\partial g\left(x_1, x_2\right)}{\partial x_2}\right]\begin{bmatrix} \dot{x}_1 \\ \dot{x}_2 \end{bmatrix}$$

$$= -l\left(x_1, x_2\right)\hat{d} + l\left(x_1, x_2\right)\left(\frac{-\dot{x}_2 - p_1 x_2 + p_1 \tau_m}{p_3}\right) + l\left(x_1, x_2\right)\cdot\frac{\dot{x}_2}{p_3} \qquad (2.29)$$

$$= -l\left(x_1, x_2\right)\hat{d} + l\left(x_1, x_2\right)\left(\frac{-p_1 x_2 + p_2 \tau_m}{p_3}\right) \quad .$$

$$= -l\left(x_1, x_2\right)\left(q + q\left(x_1, x_2\right)\right) + l\left(x_1, x_2\right)\left(\frac{-p_1 x_2 + p_2 \tau_m}{p_3}\right)$$

It can be found that Eqn. (2.29) which represents the dynamic characteristics of q(t) only involves angle and angular velocity, but does not contain the angular acceleration term, thus avoiding the problem that the acceleration signal cannot be obtained existing in Eqn. (2.25).

More generally, $l(x_1, x_2)$ can be directly defined as the differential of $g(x_1, x_2)$ with respect to the vector x, thus NDO for the SBW system can be established as

$$
\left\{
\begin{array}{l}
\dot{q} = -l\left(x_1, x_2\right)\left[-p_1 x_2 + p_2 \tau_m - p_3 g\left(x_1, x_2\right) - p_3 q\right] \\[2mm]
\hat{d}(t) = q(t) + g\left(x_1, x_2\right) \\[2mm]
l\left(x_1, x_2\right) = \dfrac{\partial g\left(x_1, x_2\right)}{\partial x} \\[2mm]
g\left(x_1, x_2\right) = K_{NDO}\, \ell_{-p_1 x_2}\left[f(y)\right]
\end{array}
\right.
\tag{2.30}
$$

where ℓ is the Lie differential operator, f(y) is the nonlinear combination of the output signals of the SBW system, $K_{NDO} = [k_1, k_2]_T$ is the gain of $g(x1, x2)$ to be designed.

Now let's discuss the stability and convergence of the NDO. Define the error of the NDO estimation as eNDO. Assume that the external disturbance changes slowly relative to the algorithm, that is, the external disturbance can be approximately regarded as an invariant within an estimation period, then the dynamic characteristics of eNDO can be described as

$$
\begin{aligned}
\dot{e}_{NDO}(t) &= \dot{\hat{d}}(t) - \dot{d}(t) \\[2mm]
&= \dot{q}(t) + \frac{d}{dt}\dot{g}\left(x_1, x_2\right) - \dot{d} \\[2mm]
&= -l\left(x_1, x_2\right)\left[-p_1 x_2 + p_2 \tau_m - p_3 g\left(x_1, x_2\right) - p_3 q\right] + \frac{\partial g\left(x_1, x_2\right)}{\partial x}\dot{x} \\[2mm]
&= -l\left(x_1, x_2\right)\left[-p_1 x_2 + p_2 \tau_m - p_3 \hat{d}(t)\right] + l\left(x_1, x_2\right)\left[-p_1 x_2 + p_2 \tau_m - p_3 d\right] \\[2mm]
&= l\left(x_1, x_2\right) p_3 \hat{d}(t) - l\left(x_1, x_2\right) p_3 d \\[2mm]
&= l\left(x_1, x_2\right) p_3 e_{NDO}
\end{aligned}
\tag{2.31}
$$

Therefore, the estimation error of system (2.31) can always be asymptotically converged to zero by means of a well-designed $l(x_1, x_2)$. The superiorities of NDO over LDO are also clearly demonstrated. Specifically, both nonlinear structure and feedback gain of function $l(x_1, x_2)$ can be flexibly designed, thus the optimal nonlinear function can be worked out offline based on the previous road spectrum data and the operating characteristics of SBW system. This enables NDO to achieve a good estimation of external rack force in all working conditions. The schematic diagram of the NDO is shown in Figure 2.17.

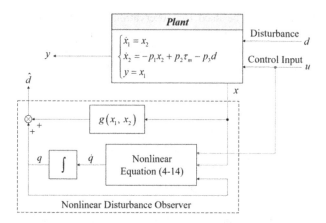

FIGURE 2.17   Structure of Nonlinear Disturbance Observer.
Source: Figure created by authors.

Compared with LDO, NDO is able to obtain a better estimation of external disturbances. The essential factor of that is d(t) itself is a nonlinear function of time, so the NDO can achieve better tracking performance. However, in the real-world automobile industry, the application of NDO is still limited due to the complexity of the theory, the lack of brevity in the algorithm, and the heavy workload of calculation.

2.4.1.3   Extended Disturbance Observer

In order to solve the problems encountered by the above two observers in practice, we introduce the concept of EDO for the SBW system based on extended state observer. The EDO absorbs the advantages of both LDO and NDO. It introduces a nonlinear gain function on the basis of the original state space observer of the system. The structure of EDO is simple and clear, the convergence rate is accelerated and the estimated performance is improved. Moreover, EDO is also convenient for engineering implementation.

The establishment of EDO starts from the LDO structure as shown in Eqn. (2.24). If we temporarily ignore the influence of external disturbance, the estimation equations representing the dynamics of the original system can be obtained as

$$\begin{cases} \dot{\hat{x}}_1 = \hat{x} + l_{LDO\_1}\left(x_1, x_2\right)e \\ \dot{\hat{x}}_2 = -p_1\hat{x}_2 + p_2\tau_m + l_{LDO\_2}\left(x_1, x_2\right)e \\ e = x_1 - \hat{x}_1 \end{cases} \tag{2.32}$$

where l_{LDO_1} and l_{LDO_2} are linear functions of the LDO system.

The above equations artificially ignore the external disturbance d(t), but d(t) indeed exists in the real system and plays a role in the estimation system. Hence the system (2.32) cannot be used for observation. Although the linear function l_{LDO} cannot restrain the external disturbance, according to theories of nonlinear feedback effect or dual system feedback effect, well-designed nonlinear feedback structure $\beta_1 h_1(e)$, $\beta_2 h_2(e)$ can effectively suppress the impact from d(t). Then, the system becomes

$$\begin{cases} \dot{\hat{x}}_1 = \hat{x}_2 + \beta_1 h_1(e) \\ \dot{\hat{x}}_2 = p_2 \tau_m + \beta_2 h_2(e) \end{cases} \qquad (2.33)$$
$$s.t.\, eh_1(e) \le 0, eh_2(e) \le 0$$

where β_1, β_2 are appropriate gain parameters; $h_1(e)$, $h_2(e)$ are appropriate nonlinear functions yielding to the constraints. When the gain parameters and the feedback functions are properly set up, for example, $h_1(e) = e$, $h_2(e) = |e|0.5\mathrm{sgn}(e)$, the observation system can achieve a good estimate of the state variables in the SBW system (2.20). Most importantly, the state observer has the ability to estimate the angle signal x_1 and the angular velocity signal x_2 without involving the external disturbance d(t), that is, not constrained by the mathematical structure of d(t), and the estimation process is performed independently of the system dynamics. The observation system (2.33) exhibits its good estimation ability, which brings great convenience to the design of EDO in the following part.

Now considering d(t) as an extended state, we define the system's extended state variable $x_3(t) = d(t)$. We also assume that the first-order differential of d(t) is bounded and satisfies $x_3 \le \zeta$. Similar to system (2.33), the EDO for the SBW system is established as shown in Eqn. (2.34), and its schematic diagram is shown in Figure 2.18.

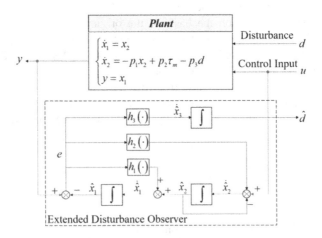

FIGURE 2.18   Structure of Extended Disturbance Observer.
Source: Figure created by authors.

$$\begin{cases} \dot{\hat{x}}_1 = \hat{x}_2 + \beta_1 h_1(e) \\ \dot{\hat{x}}_2 = -p_1 \hat{x}_2 + p_2 \tau_m - p_3 \hat{x}_3 + \beta_2 h_2(e) \\ \dot{\hat{x}}_3 = \beta_3 h_3(e) \end{cases} \quad (2.34)$$

where β_3 is the gain parameter for extended state variable estimation, $h_3(e)$ is the corresponding nonlinear function satisfying the constraint $eh_3(e) \leq 0$.

Although the design method of $h(e)$ has been given in the literature as $h(\cdot) = fal(\cdot)$, the structure is not simple, and the code implementation after necessary discretization is more complicated. Therefore, considering the simplicity of nonlinear functions, to ensure the stability of the estimation system, we set up the nonlinear functions as follows

$$\begin{cases} h_1(e) = \dfrac{e}{\delta}, h_2(e_1) = \dfrac{e}{\delta^2}, h_1(e_1) = \dfrac{e}{\delta^3} \\ \delta = \kappa |e|^\alpha \end{cases} \quad (2.35)$$

where α and κ are coefficients to be designed in order to regulate the magnitude of δ.

Next, we will discuss the stability and convergence of the EDO. Define the error vector as Γ

$$\Gamma = \begin{bmatrix} \gamma_1 & \gamma_2 & \gamma_3 \end{bmatrix}^T = \begin{bmatrix} \dfrac{x_1 - \hat{x}_1}{\delta^2} & \dfrac{x_2 - \hat{x}_2}{\delta} & x_3 - \hat{x}_3 \end{bmatrix}^T \quad (2.36)$$

Rewrite Eqn. (2.34), Eqn. (2.35) and Eqn. (2.36) as

$$(a).\delta \dot{\gamma}_1 = \frac{\dot{x}_1 - \dot{\hat{x}}_1}{\delta} = \frac{1}{\delta} \left[(x_2 - \hat{x}_2) + \frac{\beta_1}{\delta}(x_1 - \hat{x}_1) \right] \quad (2.37)$$

Eqn. (2.37) is the dynamical equation for the error estimation system of the EDO, which can be organized into a matrix expression

$$\delta \dot{\Gamma} = A_e \Gamma + \delta B_e \dot{d} \quad (2.38)$$

where $A_e = \begin{bmatrix} -\beta_1 & 1 & 0 \\ -\beta_2 & -p_1\delta & -p_3 \\ -\beta_3 & 0 & 0 \end{bmatrix}$ is the system matrix of EDO estimation error; $B_e = [0\ 0\ 1]_T$ is the excitation matrix.

Since the system parameters p_1 and p_3 are known and bounded, when feedback gains β_1, β_2, β_3, and δ are selected wisely, A_e can always satisfy the Hurwitz condition. Therefore, for any given positive definite matrix X, another positive definite matrix Q can always be found satisfying the following Lyapunov equation

$$A_e^T X + XA_e = -Q \tag{2.39}$$

Define the Lyapunov function of the EDO as $V_{EDO} = \delta\Gamma_T X\Gamma$, then

$$
\begin{aligned}
\dot{V}_{EOD} &= \delta\dot{\Gamma}^T X\Gamma + \delta\Gamma^T X\dot{\Gamma} \\
&= \left(A_e\Gamma + \delta B_e\dot{d}\right)^T X\Gamma + \Gamma^T X\left(A_e\Gamma + \delta B_e\dot{d}\right) \\
&= \Gamma^T\left(A_e^T X + XA_e\right)\Gamma + 2\delta\Gamma^T\left(XB_e\right)\Gamma \\
&\leq -\Gamma^T Q\Gamma + 2\delta\left\|\dot{d}\right\|\cdot\left\|XB_e\right\|\cdot\left\|\Gamma\right\| \\
&\leq -\lambda_{min}\left(Q\right)\left\|\Gamma\right\|^2 + 2\delta\zeta\left\|XB_e\right\|\cdot\left\|\Gamma\right\|V
\end{aligned}
\tag{2.40}
$$

where $\lambda_{min}(Q)$ is the minimum eigenvalue of the positive definite matrix Q, $\|\cdot\|$ is the Euclidean norm.

According to Eqn. (2.40), the stability condition of the EDO can be obtained

$$\left\|\Gamma\right\| \leq \frac{2\delta\xi\left\|XB_e\right\|}{-\lambda_{min}\left(Q\right)} \tag{2.41}$$

By adjusting α and κ, δ can always be obtained to meet the above condition. From Eqn. (2.40) and Eqn. (2.41), it can be found that the convergence speed of the error vector Γ is related to δ. According to the theory of singularly perturbed system, the error dynamic Eqn. (2.38) is the fast-changing subsystem of the actual SBW system, and $\|\Gamma\|$ is a high-order infinitesimal with respect to δ. As δ decreases, the error gradually tends to zero. Besides, the smaller δ is, the faster the convergence speed will be. The essential reason why EDO can well estimate the external disturbance d(t) is that as long as the system is observable, no matter what mathematical form d(t) has, once it acts on the SBW system, it will inevitably generate a corresponding response to the output of the system. Hence, the key is to extract the response from the output information and recover it in real time. The high estimation efficiency of EDO during the recovery process is mainly due to the design of its nonlinear gain function.

In addition, it can be seen from Eqn. (2.34) that the inputs of EDO are the motor torque τm and the angle signal x_1, without the angular velocity signal x_2 or other state information of the system, which is consistent with the actual SBW execution hardware and sensor interface. Compared with other DOs, EDO not only has fewer input variables, but also has the advantages of simple structure and fast convergence, which is very important for steering-following control that emphasizes real-time performance. Meanwhile, the velocity signal estimated by the EDO can be compared with that output by traditional differential algorithm or the fastest tracking differentiator (TD), so as to further verify the effectiveness of EDO.

2.4.2 Design of Parameter Adaptive Placement Controller

We define θ_{ref} as the steering angle input or the angle reference signal indicating the driver's steering intention which is sent by the handwheel module of the SBW system. Note that θ_{ref} here is not the direct output signal of the sensor, but the optimal signal synthetically refined by the upper layer algorithm based on the angle input, the torque input, the vehicle dynamics and the environmental factors. Thus, θ_{ref} can be assumed to be a smooth signal.

The objective of the steering-following algorithm is that the actual angle output θ of the execution system should be able to follow the target angle θ_{ref} accurately and fast even if the external disturbance exists and changes drastically. In the field of control, this kind of problem is called static error-free tracking problem. That is, when external disturbance exists in the system, for any non-zero θ_{ref}, there is always control input u such that

$$\lim_{t \to \infty} \theta(t) = \lim_{t \to \infty} \theta_{ref}(t) \tag{2.42}$$

The static error-free controller usually includes two main parts: the servo controller and the stabilizing regulator. The servo controller is usually a Linear Time Invariant (LTI) system, such as a common PID controller, which is used to realize the asymptotic tracking and disturbance suppression. The stabilizing regulator is used to configure the pole distribution of the plant, so that all poles are located on the left-hand side of the complex plane, to ensure the stability of the system. State feedback is a commonly used method for stabilizing regulation. Based on the concept of the static error-free controller, here we propose Parameter Adaptive Placement Controller (PAPC) to deal with the time-variant parameters of the SBW execution system.

First of all, the servo controller can be described by a set of linear differential equations since it is an LTI system. We define x_c as the state vector of the servo controller. The corresponding system matrix is A_c and the input matrix is B_c. Benefiting from the good speed-estimation property of the fastest TD, the input vector e of the servo controller can include errors of angle and angular velocity at the same time. The output u is generally designed from x_c. Then the dynamic equations of the servo controller can be expressed as

$$\begin{cases} \dot{x}_c = A_c x_c + B_c e \\ u_c = -K_c x_c \end{cases} \tag{2.43}$$

The stabilizing regulator is a feedback link in the control loop of the system. It can be divided into two parts: output feedback and state feedback, and the latter has greater freedom of design. For SBW system (2.20), all states are observable, hence we use state feedback to stabilize the system. Define K as the state feedback matrix, then the dynamic equations of the system including the stabilizing regulator can be described as

$$\begin{cases} \dot{x} = Ax + Bu + Ed \\ \quad u_s = -Kx \end{cases} \qquad (2.44)$$

Combining Eqn. (2.43) and Eqn. (2.44), the static error-free controller of SBW system can be obtained as

$$\begin{cases} \begin{bmatrix} \dot{x} \\ \dot{x}_c \end{bmatrix} = \begin{bmatrix} A & 0 \\ -B_c C & A_c \end{bmatrix} \begin{bmatrix} x \\ x_c \end{bmatrix} + \begin{bmatrix} B \\ 0 \end{bmatrix} u + \begin{bmatrix} E \\ 0 \end{bmatrix} d + \begin{bmatrix} 0 \\ B_c \end{bmatrix} x_r \\ \quad u = u_s + u_c = \begin{bmatrix} -K & K_c \end{bmatrix} \begin{bmatrix} x \\ x_c \end{bmatrix} \end{cases} \qquad (2.45)$$

where x_r is the reference angle signal. The above equation is the standard form of the static error-free controller.

To facilitate the design of subsequent control algorithms as well as the analysis of system performance, Eqn. (2.45) can be rewritten as

$$\begin{aligned} \begin{bmatrix} \dot{x} \\ \dot{x}_c \end{bmatrix} &= \begin{bmatrix} A & 0 \\ -B_c C & A_c \end{bmatrix} \begin{bmatrix} x \\ x_c \end{bmatrix} + \begin{bmatrix} B \\ 0 \end{bmatrix} \begin{bmatrix} -K & K_c \end{bmatrix} \begin{bmatrix} x \\ x_c \end{bmatrix} + \begin{bmatrix} E \\ 0 \end{bmatrix} d + \begin{bmatrix} 0 \\ B_c \end{bmatrix} x_r \\ &= \begin{bmatrix} A - BK & BK_c \\ -B_c C & A_c \end{bmatrix} \begin{bmatrix} x \\ x_c \end{bmatrix} + \begin{bmatrix} E \\ 0 \end{bmatrix} d + \begin{bmatrix} 0 \\ B_c \end{bmatrix} x_r \end{aligned} \qquad (2.46)$$

where A_{PAPC} is the system matrix connecting the static error-free controller and the SBW system in series.

According to Eqn. (2.46), the standard static error-free controller essentially realizes the pole configuration of A_{PAPC} by designing the gain matrix $[-K, -Kc]$. In engineering practice, given a set of expected poles λ^*, the characteristic equation of A_{PAPC} can be found inversely, and the parameters of the controller can be worked out by matching the corresponding polynomial coefficients.

It can be found that the stabilizer design or pole configuration process of the standard static error-free controller is carried out for the LTI system. Except for the parameters $[k_1, k_2, kc_1, kc_2]$ to be designed, the rest of the elements in A_{PAPC} are constant. Since the model parameters of the SBW system are time-variant, the constant controller parameters cannot guarantee that the poles of A_{PAPC} are all configured at their desired positions. Due to the strong real-time property and online estimation ability of the state-space identification method, here we design the PAPC, which can adaptively adjust the control parameters $[k_1, k_2, kc_1, kc_2]$ by online estimation of system parameters, to ensure that the poles of system matrix are always located near the desired poles λ^*. In practice, the loaded SBW system requires the online identification method to consider the effect of external disturbance d(t). Here, the identification algorithm can substitute d(t) by the output of disturbance observer $\hat{d}$, and consider the sum of the external disturbance and the motor torque as the equivalent excitation of the identification system. Based on that, the original state space identification algorithm can be revised as

$$
\dot{\hat{p}}_1 = \begin{cases} -\gamma_1 \hat{x}_2 \varepsilon & p_{1L} \leq \hat{p}_1 \leq p_{1U} \\ 0 & otherwise \end{cases}, \hat{p}_1(0) = \hat{p}_{10}
$$

$$
\dot{\hat{p}}_2 = \begin{cases} \gamma_2 u_{eq} \varepsilon & p_{2L} \leq \hat{p}_2 \leq p_{2U} \\ 0 & otherwise \end{cases}, \hat{p}_2(0) = \hat{p}_{20} \tag{2.47}
$$

$$
u_{eq} = \tau_m + i_{mc}\hat{d}
$$

where u_{eq} is the equivalent excitation considering the motor input and the external disturbance.

Next, let's design the PAPC. The servo controller is designed as

$$
\begin{cases} \dot{x}_c = \underset{A_c}{[0]} x_c + \underset{B_c}{[1 \quad 0]} \begin{bmatrix} e \\ \dot{e} \end{bmatrix} \\ u_c = k_{c1} x_c \end{cases} \tag{2.48}
$$

Meanwhile, we design the feedback matrix of the stabilizing regulator as $K = [k_1, k_2]$, then the dynamic equations of the system can be represented as

$$
\begin{bmatrix} \dot{x}_1 \\ \dot{x}_2 \\ \dot{x}_x \end{bmatrix} = \begin{bmatrix} 0 & 1 & 0 \\ 0 & -\hat{p}_1 & 0 \\ -1 & -k_{c2} & 0 \end{bmatrix} \begin{bmatrix} x_1 \\ x_2 \\ x_c \end{bmatrix} + \begin{bmatrix} 0 \\ \hat{p}_2 \\ 0 \end{bmatrix} \underbrace{[-k_1 \quad -k_2 \quad k_{c1}]}_{u} \begin{bmatrix} x_1 \\ x_2 \\ x_c \end{bmatrix} + \begin{bmatrix} 0 \\ -\hat{p}_3 \\ 0 \end{bmatrix} \hat{d} + \begin{bmatrix} 0 & 0 \\ 0 & 0 \\ 1 & k_{c2} \end{bmatrix} \begin{bmatrix} x_{r1} \\ x_{r2} \\ x_r \end{bmatrix}
$$

$$
= \underbrace{\begin{bmatrix} 0 & 1 & 0 \\ -\hat{p}_2 k_1 & -\hat{p}_1 - \hat{p}_2 k_2 & -\hat{p}_2 k_{c1} \\ -1 & -k_{c2} & 0 \end{bmatrix}}_{A_{PAPC}} \begin{bmatrix} x_1 \\ x_2 \\ x_c \end{bmatrix} + \begin{bmatrix} 0 \\ -\hat{p}_3 \\ 0 \end{bmatrix} \hat{d} + \begin{bmatrix} 0 & 0 \\ 0 & 0 \\ 1 & k_{c2} \end{bmatrix} \begin{bmatrix} x_{r1} \\ x_{r2} \end{bmatrix} \tag{2.49}
$$

According to Eqn. (2.49), the control parameters $[k_1, k_2, k_{c1}, k_{c2}]$ can always be designed so that the poles of A_{PAPC} are λ^*. In other words, the control system is stable when all λ^* lie to the left-hand side of the complex plane. Moreover, the larger negative real part λ^* has, that is, the farther the position of poles is from the imaginary axis, the faster the system converges, and the faster the actual angle θ follows the target θ_{ref}. The structure of PAPC is shown in Figure 2.19.

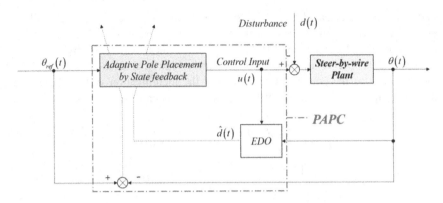

FIGURE 2.19   Structure of Parameter Adaptive Placement Control.
Source: Figure created by authors.

Next, we will further analyze the actual physical meaning of the control system. Organize the control input u of the SBW execution system into the following form

$$
\begin{aligned}
u &= -k_1 x_1 - k_2 x_2 + k_{c1} x_c \\
&= -k_1 \theta - k_2 \dot{\theta} + k_{c1} \left(k_{c2} e + \int e\, dt \right) \\
&= \underbrace{-k_1 \theta - k_2 \dot{\theta}}_{\text{State Feedback}} + \underbrace{k_p e + k_i \cdot \int e\, dt}_{\text{PI}}
\end{aligned}
\tag{2.50}
$$

It can be seen that the control input consists of two parts. The PI controller corresponds to the servo controller in the algorithm, which is used to suppress the adverse effects of the external disturbance d(t) on the tracking control. The state feedback part uses the angle and angular velocity of the actual system for stabilizing regulation or pole configuration. Although its mathematical structure is similar to feedforward compensation, it is contrarily the feedback mechanism to keep the system stable.

In conclusion, PAPC adaptively adjusts the controller parameters according to the changes of parameters in the system model, which effectively suppresses the influence of the model parameter uncertainty on the control accuracy, improves the dynamic response speed of the system, and improves the tracking performance to some extent. However, for SBW system, the change of external disturbance is the main factor affecting the dynamic characteristics of the system relative to the change of model parameters. According to the previous analysis, PAPC suppresses the random and unknown disturbance only by the integral link in the traditional PI controller, which will inevitably lead to lags in the dynamic response of the system.

2.4.3 Design of Active Tracking Controller for SBW System

The PAPC proposed in the previous section mainly solves the problem of time-variant model parameters in the actual working process of SBW system. It improves the steering-following performance by overcoming the uncertainty of model parameters. However, compared with model uncertainty, the influence of external disturbance on the system dynamics is more prominent. In order to obtain good dynamic response characteristics, the usage of the integral term in PI regulator to suppress interference is far from enough. Therefore, it is very important to design a controller with disturbance immunity by itself.

Similar to the concept of an active suspension system, if the external disturbance can be correctly estimated and actively compensated, the following performance of SBW system will be greatly improved. At the same time, although the impact of parameter uncertainty is less than that of external disturbance, the parameter uncertainty or model uncertainty need to be suppressed in the real-world control process. Consequently, we address this issue by designing a controller with GRF suppression capability. Note that the GRF is obtained by the disturbance observer, hence it is equivalent with external disturbance in terms of text description. Based on that, we propose the Active Tracking Controller (ATC), a tracker suitable for the SBW system, to handle the steering-following task. The ATC is derived from the robust sliding mode controller which combines the disturbance observer. It uses $\hat{d}$, the output of EDO, to actively perform feed-forward compensation on the sliding mode controller,

As mentioned above, the steering tracking error is defined as $e = \theta_{ref} - \theta$. Combined with Eqn. (2.18), the dynamic equation of the tracking error can be obtained as

$$\ddot{e} = \ddot{\theta}_{ref} - \ddot{\theta} = \ddot{\theta}_{ref} + p_1\dot{\theta} - p_2\tau_m + p_3 d \qquad (2.51)$$

Note that θ_{ref} here is a smooth signal defined by the upper-level algorithm, so its second-order differential has practical significance. Meanwhile, the model parameters in Eqn. (2.51) are all fixed as appropriate nominal parameters. Parameter uncertainty and model uncertainty of the system are included in the disturbance term.

The sliding mode variable is defined as

$$s = \dot{e} + \wedge e \qquad (2.52)$$

where $\wedge > 0$ is the sliding mode parameter to be designed.

Differentiate the sliding mode variable s and combine it with Eqn. (2.52) to get

$$\dot{s} = \ddot{e} + \wedge\dot{e}$$
$$= \ddot{\theta}_d + \wedge\dot{e} + p_1\dot{\theta} - p_2\tau_m + p_3 d \qquad (2.53)$$

The above equation shows the dynamic characteristics of s. Compared with Eqn. (2.51), it reduces the order of the equation, which is helpful for the derivation and the stability proof of the subsequent ATC.

Furthermore, rearrange Eqn. (2.53) into the following matrix expression form

$$\dot{s} = p_2 \left(\begin{bmatrix} \dfrac{1}{p_2} & \dfrac{p_1}{p_2} & \dfrac{p_3}{p_2} \end{bmatrix} \begin{bmatrix} \ddot{\theta}_d + \wedge \dot{e} \\ \ddot{\theta} \\ d \end{bmatrix} - \tau_m \right) \tag{2.54}$$
$$= p_2 \left[W^T \phi(x) - \tau_m \right]$$

Theoretically, to make the sliding mode variable asymptotically approach zero, the input of the controller should be designed as

$$\tau_m = W^T \phi(x) \tag{2.55}$$

Eqn. (2.55) is an ideal feed-forward controller. However, since the model (2.18) is a simplified mathematical model based on several assumptions, and the SBW system is seriously disturbed by noise and stimulated by high-order nonlinear items in the application, the simple feedforward control link is insufficient. An error feedback loop needs to be designed at the same time to suppress the impact of these unfavorable factors. Thus, the ATC is designed as

$$\tau_m = W^T \phi(\hat{x}) + k_s s \tag{2.56}$$

where $\hat{x}$ is the estimated value of corresponding state variable; k_s is the feedback gain to be designed; $\hat{d}$ is the estimated result of EDO.

According to Eqn. (2.56), $k_s s$ is actually a PD regulator. The existence of the fastest TD ensures the rationality of the design, because TD does not depend on the dynamic model of the system, but can realize fast and accurate estimation on the angular velocity only according to the angle signal. The algorithm structure of ATC is shown in Figure 2.20.

Here the stability and convergence of the ATC algorithm will also be discussed. Define VATC as the Lyapunov function of ATC as shown in Eqn. (2.57)

$$V_{ATC} = \frac{1}{2} s^2 \tag{2.57}$$

Then we can obtain $\dot{V}_{ATC}$ as

$$\dot{V}_{ATC} = s\dot{s} = p_2 s \left[W^T \left(\phi(x) - \phi(\hat{x}) \right) - k_s s \right] \tag{2.58}$$
$$= p_2 \left[-k_s s^2 + s W^T \tilde{\phi}(x) \right]$$

where $\tilde{\phi}(x)$ is the estimation error matrix.

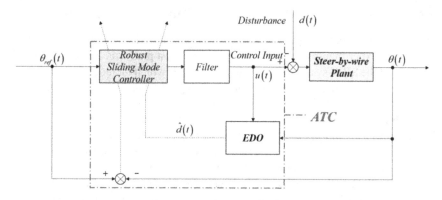

FIGURE 2.20   Structure of Active Tracking Controller.

Source: Figure created by authors.

Since p1, p2 and $\hat{d}$ are all bounded, the error term $W^T \tilde{\phi}(x) \le \Delta$ is also bounded, then Eqn. (2.58) can be expressed in the inequality form as

$$\dot{V}_{ATC} \le p_2 \left[-k_s s^2 + \frac{1}{2}(s^2 + \Delta^2) \right]$$
$$= -p_2 \left[(2k_s - 1)\frac{s^2}{2} - \frac{1}{2}\Delta^2 \right] \qquad (2.59)$$
$$= -p_2 \left[(2k_s - 1)V_{ATC} - \frac{1}{2}\Delta^2 \right]$$

The solution of (2.59) is shown as

$$V_{ATC}(t) \le e^{-p_2(2k_s - 1)(t - t_0)}V_{ATC}(t_0) + \frac{p_2 \Delta^2}{2} \int_{t_0}^{t} e^{-p_2(2k_s - 1)(t - \tau)}d\tau$$
$$= e^{-p_2(2k_s - 1)(t - t_0)}V_{ATC}(t_0) + \frac{\Delta^2}{4k_s - 2}\left(1 - e^{-p_2(2k_s - 1)(t - t_0)}\right) \qquad (2.60)$$

As $k_s > 0.5$,

$$\lim_{t \to +\infty} V_{ATC}(t) \le \frac{\Delta^2}{4k_s - 2} = \bar{\upsilon}_{ATC} \qquad (2.61)$$

As $t \to +\infty$, the Lyapunov function of the ATC converges exponentially to $\bar{\upsilon}_{ATC}$. The convergence speed is decided by the feedback gain k_s. Larger k_s always leads to faster convergence speed.

2.5   VERIFICATION OF ACTIVE TRACKING CONTROL

In order to evaluate the performance of the steer-by-wire system, hardware in loop (HIL) experiments have been conducted on a test platform. The experimental platform of the SBW system is shown in Figure 2.21, which includes the handwheel module, steering actuator module, and steering resistance module. A column electric power steering systems (CEPS) is applied as the upper steering handwheel part, while a pinion electric power steering system (PEPS) is employed as the lower steering actuation part. The steering resistance module is made by a electric servo cylinder loading system, which can simulate the resisting force during steering motion. The parameters specification of the steering actuating system is the same as Table 2.1 in section 2.3.3. There is no mechanical connection between the steering column and the front steering wheel. The steering control designed that follows the driver input is realized by the signal transmission between the electric control units. A rapid control prototype (dSPACE PX20) is used as the central controller, and the algorithm model is built under Matlab/Simulink environment. The sampling rate of system is selected as 10 ms. CAN bus communication is applied between the control units.

In order to verify the effectiveness of the ATC approach in different steering condition, not only regular test cycle but also critical ones must be taken into consideration. Based on the ISO standards for vehicles, the test conditions for steering control are designed as follows: Weave Test and Transition Test are two basic steering tests with relative small rack force; The Square Turn and Serpentine Test are two critical steering tests, with high steering velocity and large disturbance; the Belgian Test is a special steering condition with a high frequency vibration of

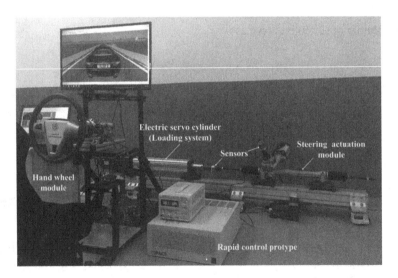

FIGURE 2.21   Experimental platform of SBW system.

Source: Figure created by authors.

TABLE 2.6
Characteristic Values of Test Cycles

Maneuver Parameters	Weave Test	Transition Test	Square Turn	Serpentine	Belgian
Max.Steering Angle (deg)	40	45	540	540	440
Max.Steering Velocity (deg/s)	10	5	800	100	50
Max.Rack Force (N)	1000	800	9000	5600	3750
Max.Rack Force Gradient (N/s)	640	300	1.7e5	1.2e5	1e5
Vehicle Speed in Operation (km/h)	100	100	25	10	40

Source: Table created by authors.

the rack force. The main characteristic values of the test cycles mentioned above are listed in Table 2.6 [8].

Based on the analysis in the above section, the rack force and system friction are taken as a lumped disturbance during the design of observer. The estimated rack force is obtained by extracting the friction term from the observer output. To simplify the modeling, we regard the system friction as a coulomb friction, which is represented by a sign function related with velocity. The experimental results of rack force estimation in different cycles are illustrated in Figures 2.22–2.26.

To show the improvement of the proposed ATC control approach, a parameter adaptive feedforward controller (PAFC) designed in the previous research has been utilized in the experiments as a comparison. The experimental results of tracking performance in different steering test cycles are shown in Figures 2.27–2.31, where the referenced steering wheel angle of Weave Test (Figure 2.27) Transition Test (Figure 2.28), Square Turn (Figure 2.29), Serpentine Test (Figure 2.30) and Belgian Test (Figure 2.31) maneuver are generated by computer according to ISO standards.

The experimental studies cover most of steering conditions to verify the proposed SBW controller. The weave test and transition test are the typical vehicles on-center handling operation to represent most of the realistic driving conditions. The square turn and serpentine tests stand for emergency driving conditions, which have high value rack force and fast varying gradient. The Belgian test is a special steering condition, in which the vehicle passes a gravel road with high frequency vibration. From the results of Figures 2.27–2.31, the steering performance in these driving conditions show the effectiveness of proposed SBW controller.

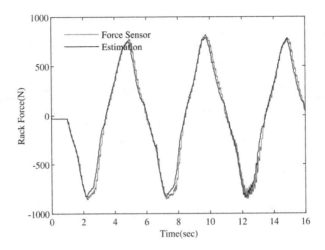

FIGURE 2.22   Rack force estimation of weave test maneuver.
Source: Figure created by authors.

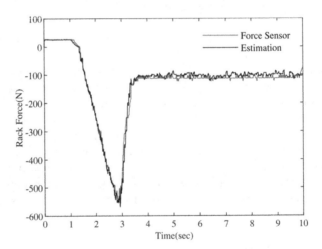

FIGURE 2.23   Rack force estimation of transition maneuver.
Source: Figure created by authors.

From the experimental results in Figure 2.22 and Figure 2.23, the rack force estimation is close to the measured values, which means the second-order model gives desired fidelity for the rack force estimation in the test cycle. From Figure 2.24 and Figure 2.25, the estimation errors become larger at the point of flexion. This is because when the actuator operates at high velocity with large load, the unmodelled high order components of the SBW system will increase the dynamic error. From

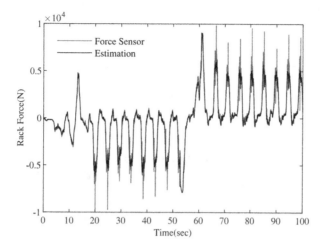

FIGURE 2.24   Rack force estimation of square turn maneuver.

Source: Figure created by authors.

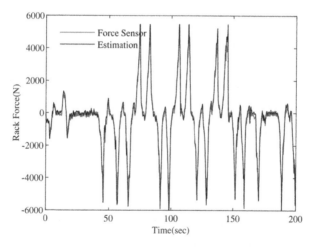

FIGURE 2.25   Rack force estimation of serpentine maneuver.

Source: Figure created by authors.

the results in Figure 2.26, despite the high frequency vibration, it still achieved desired estimation of the rack force due to the large bandwidth of the EDO.

Figures 2.27–2.31 have illustrated the tracking performance of the ATC method. The PAFC method is a modified PID controller with adaptive law to avoid parameter disturbances. It can be found that the steering performance of the ATC has achieved better tracking results. The robustness of the SBW steering control

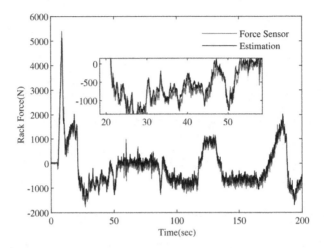

FIGURE 2.26   Rack force estimation of Belgian maneuver.

Source: Figure created by authors.

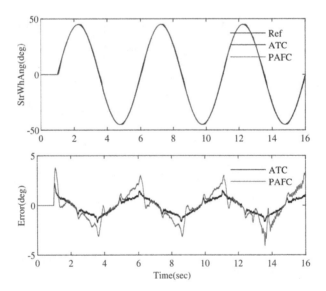

FIGURE 2.27   Tracking performance of weave test maneuver.

Source: Figure created by authors.

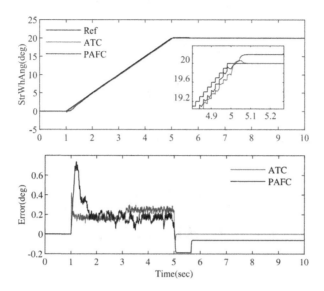

FIGURE 2.28   Tracking performance of transition maneuver.

Source: Figure created by authors.

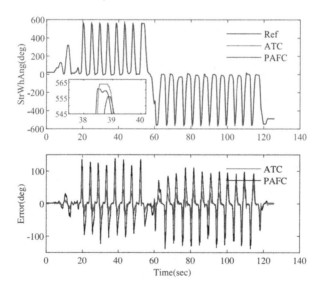

FIGURE 2.29   Tracking performance of square turn maneuver.

Source: Figure created by authors.

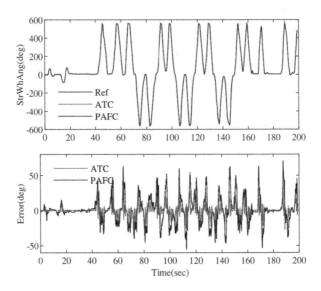

FIGURE 2.30   Tracking performance of serpentine maneuver.

Source: Figure created by authors.

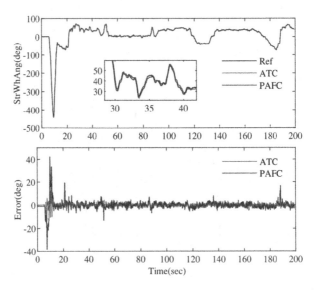

FIGURE 2.31   Tracking performance of Belgian maneuver.

Source: Figure created by authors.

is improved by the disturbance compensation, especially in the critical driving conditions with high steering velocity and large road disturbance. It should be noted that the steering actuator cannot respond to disturbance compensation with high frequency, which limit the improvement of the experimental results.

2.6 CONCLUSION

In this chapter, a concept of an active tracking controller is proposed to improve the tracking performance of SBW system. Compared with different disturbance observers, an extended disturbance observer is designed for the rack force estimation. The proposed observer shows better estimation accuracy and control robustness both in SIL and HIL experiments. Based on the disturbance estimation, an active disturbance compensation approach is designed with the slide mode controller for SBW system. Compared with a previous PID-based steering controller, the proposed ATC method achieved better tracking control in the steering test cycles, even in some critical driving conditions with high steering velocity and large external load.

REFERENCES

[1] Wang, H., Man, Z., Shen, W., et al. (2014). Robust control for steer-by-wire systems with partially known dynamics. *IEEE Trans. Indus. Inform.*, 10(4):2003–2015.

[2] Wu, X., Ye, C., Xu, M. (2016). Two-port network based bilateral control of a steer-by-wire system. *Int. J. Automo. Technol.*, 17(6):983–990.

[3] Cetin, A. E., Adli, M. A., Barkana, D. E. and Kucuk, H. (2010). Implementation and development of an adaptive steering-control system. *IEEE Trans. Veh. Technol.*, 59(1):75–83.

[4] Setlur, P., Wagner, J. R., Dawson, D. M. and Braganza, D. (2006). A trajectory tracking steer-by-wire control system for ground vehicles. *IEEE Trans. Veh. Technol.*, 55(1):76–85.

[5] Wu, X., Zhang, M., Xu, M. (2018). Adaptive feedforward control of a steer-by-wire system by online parameter estimator. *Int. J. Automo. Technol.*, 19(1):159–166.

[6] Ioannou, A. P. and Sun, J. (1996). *Robust adaptive control.* Prentice Hall, Upper Saddle River, NJ, USA.

[7] Sinha, N. K. and Kuszta, B. (1983). *Modeling and identification of dynamic systems.* Springer, Berlin, Germany.

[8] Wu, X., Zhang, M. and Xu, M. (2019). Active tracking control for steer-by-wire system with disturbance observer. *IEEE Trans. Veh. Technol.*, 68(6):5483–5493.

3 Force Feedback Control for Steer-by-Wire System

3.1 INTRODUCTION

The steer-by-wire (SBW) system is an important part of the wire-controlled chassis, which is considered to be representative of the next generation of automotive steering systems. The SBW system achieves the mechanical decoupling between the steering wheel module and the actuator module, replacing them with sensors, the control unit and electromagnetic actuators, thus featuring safety and high real-time performance. It contributes to the development of an integrated chassis, allows for more space in the vehicle, and provides complete steering system hardware support for autonomous driving technology.

The SBW system eliminates the mechanical linkage between the steering column and the steering actuator – that makes it difficult for the driver to perceive road information. The steering feel from the torque feedback should to be designed to reflect the maneuvering dynamics correctly. Therefore, it is especially important to design a clear and realistic steering feel to improve driving comfort and safety. In providing this steering feel, the steering torque feedback in the steering wheel system can be generated based on either the force signal from load cells attached to the rack, or from the mathematical model designed to produce the steering feel, which is indispensably dependent on the estimation of tire parameters [1]. The future SBW system should not only be able to generate the usual steering feel like a conventional steering system, but also the steering feel should be tunable in a quite intuitive way to meet the individual driver's requests [2]. The characteristic of by-wire structure provides a possibility for the customizable steering experience, which can improve driving stability and maneuverability, as well as for autonomous steering control to assist the driver.

Various studies on the design of the steering feedback torque in the SBW system confirm the challenge of generating both a realistic and a highly customizable steering feel. Alignment torque is an important part of the steering feel. Fankem and Muller take the aligning torque as the main source of the steering feel and add the inertia, damping and friction of the system as compensation to reconstruct the steering feel [2]. How to evaluate the aligning torque is a key issue during the force feedback control for steer-by-wire system.

 DOI: 10.1201/9781003481669-3

One way to obtain alignment torque is to establish a vehicle dynamics model, combined with a nonlinear tire model to calculate the aligning moment of the vehicle [3], [4]. The accuracy of this approach depends greatly on the accuracy of the vehicle parameters and tire parameters, which are difficult to obtain in practical applications. This kind of steering feel needs a fidelity steering model with abstraction of the physical system. Low-fidelity models, typically involving a spring model based on steering angle, have been used on simulators by Segawa et al. [5] and Oh et al. [6]. Due to the low fidelity of spring models, the dynamic characteristic during steering control cannot be expressed to show the important elements of the steering feel. Further to the issue of low fidelity steering models, higher fidelity models for the steering system have been proposed by Salaani et al. [7] and Mandhata et al. [8]. Though higher fidelity models can capture the dynamic characteristic of the steering feel, the increased complexity of the model makes parameter identification and calibration become a major challenge.

Another way is to calculate the alignment torque utilizing the rack force of the steering actuator module. The rack force can reflect the actual steering operation of the vehicle. It is possible to obtain rack force by installing force sensors, which certainly increases the cost due to the expensive sensors [9]. Based on the extensive experimental data, the rack force of the steering actuation system can also be generated by a torque map coupled with steering motor current [10]. Though this technique generates a realistic steering feel, this feel is tied to the physical properties of the vehicle and cannot easily be varied in software. At present, the feasible way is to estimate the rack force by designing an observer, which is highly feasible and greatly reduces the cost. Li et al. designed a disturbance observer based on the dynamical model of steering system [11]. Zhang et al. design a distribution observer and applied Kalman filter to improve performance [12].

Based on previous study, this chapter proposes a method of reconfigurable steering feel based on a hybrid framework, which in turn is based on the decomposition of steering dynamics. This approach estimates the rack force by an extended disturbance observer to provide real-time feedback of the dynamic loads from the tire–road interaction. Compared with traditional steering feel reconstruction by forward dynamics, the system friction torque can also be employed as the composition of steering feel. Thus, a friction torque-assisting model is proposed to further optimize the smoothness of steering feel. The steering performance of this kind of hybrid framework is analyzed by the objective evaluation indicators of hardware-in-the-loop experiments.

3.2 MODELLING OF STEERING FORCE FEEDBACK SYSTEM

For conventional vehicles, electric power steering systems can generate optimized steering feel by different assistance torque. Steer-by-wire vehicles should have characteristics similar to conventional vehicles. Therefore, higher fidelity steering-feel models that can create a wide variation of steering feel are necessary. However, the increased complexity of the model makes parameter identification and calibration become major challenges. Therefore, finding the appropriate level of steering model fidelity is critical. The model must be complex enough to capture all the elements of steering feel that modern drivers care about, while remaining simple enough to be tuned intuitively [13].

3.2.1 DYNAMICS OF SBW SYSTEM

As shown in Figure 3.1, the steer-by-wire system consists of three main modules: the steering wheel module, also called the steering wheel actuator (SWA); the actuator module, also known as the steering rack actuator (SRA); and the vehicle control unit (VCU). Signals such as steering angle and torque detected by the sensor from SWA are transmitted to the vehicle control unit by CAN bus, while the VCU obtains the feedback signals by the sensor from SRA, including output torque of the actuator motor, rack displacement, and so forth. On the one hand, VCU controls the actuator motor to realize the steering process according to the desire of the driver, and on the other hand, it controls the steering feel motor to feed back the steering feel and road information to the driver.

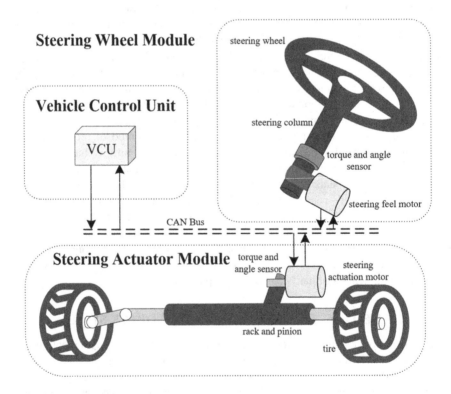

FIGURE 3.1   Schematic diagram of SBW system.

Source: Figure created by authors.

In this chapter, it is assumed that the steering modules can be regarded as a rigid transmission system [14]. The components of the steering wheel module can be equivalent to the steering column at SWA. For the steering actuator module, the rotational inertia of each component can be equivalent to the axis of the steering column at SRA. Both the steering wheel module and the actuator module can be separately described as second-order systems. A simplified steering model can be designed to capture the effect of tire torques, power assist, the inherent inertia and damping of a steer-by-wire steering system.

For the steering wheel module, there exists

$$J_{sw}\ddot{\theta}_{sw} + B_{sw}\dot{\theta}_{sw} + \sum f_{sw} + i_{ms} \cdot \tau_{ms} = T_{sw} \qquad (3.1)$$

where, θ_{sw} is the steering wheel angle, which can be collected by the sensor and sent to the CAN bus. J_{sw} is the equivalent rotational inertia of the steering wheel module. B_{sw} is the equivalent damping coefficient. $\sum f_{sw}$ is the total friction of SWA. T_{sw} represents the torque of the steering wheel, which can also be approximated as the torque felt by the driver. τ_{ms} is the output torque of the steering feel motor. i_{ms} is the reduction ratio from the steering feel motor to the steering column. The differential equation for the dynamics of the steering actuator is shown in (3.2).

$$J_m\ddot{\theta}_m + B_m\dot{\theta}_m + \tau_{fri} = i_{mc} \cdot \tau_m - F_{rack} \cdot r \qquad (3.2)$$

where, θ_m indicates the angle of the steering column shaft at SRA. J_m and B_m are the equivalent rotational inertia and equivalent damping coefficients, respectively. τ_{fri} representing the total friction of the steering actuator module. i_{mc} is the reduction ratio from the output shaft of the actuator motor to the shaft of the steering column. τ_m indicates the output torque of the actuator motor. F_{rack} is the rack force, which can be obtained from the disturbance observer. r is the radius of the pinion.

3.2.2 Design of Reconfigurable Steering Feel

The design of reconfigurable steering feel for SBW aims to provide the driver with a realistic steering feedback torque, which implies the driver should be provided with the similar functionality of a conventional steering system. Based on the study in Fankem [2], a structure of the numerical model to compute the desired steering torque is shown in Figure 3.2, which can ideally supply the driver with his preferred steering feel. The composition of the steering torque can be mainly divided into two parts: compensation torque and aligning torque. The aligning torque of a steering system is one of the major components of steering feel. The calculation of aligning torque is based on the rack force of the steering mechanism, which can be achieved by the design of disturbance state observer. The compensation torques for steering feel are composed of inertia compensation, damping compensation, friction compensation and assistance compensation. These compensation torques are calculated by the vehicle dynamics information, and state identification is

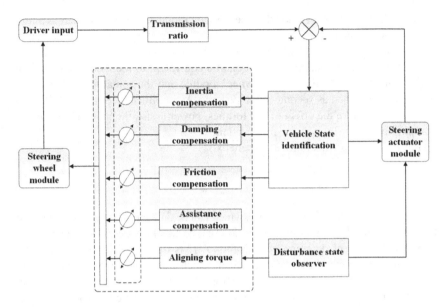

FIGURE 3.2   Design of reconfigurable steering feel.

Source: Figure created by authors.

necessary for the controller. The final steering feel is a weighted sum of these torques. This parallel structure offers the possibility to introduce tuning factors, which serve here as weights of different torque components, and could achieve a reconfigurable steering feel to meet the demands of individuation.

3.2.3   RECONFIGURABLE STEERING FEEL

From Eqn. (3.1), there are inertia, damping and friction in the steering wheel module itself. The method discussed in this chapter introduces the rack force, which is an important part of steering feel so that the driver can feel the information of the load force in real time. The rack force can be estimated by means of an extended disturbance observer instead of a force sensor, which greatly reduces costs. The introduction of a friction torque assisting model employs system friction torque as the composition of steering feel, which contributes to further optimization of the steering feel. Additional damping compensation is introduced to ensure that the return process is not overshooting. The limiting torque helps the driver to perceive the limit position when the vehicle is stationary.

According to the source analysis of the steering feel, the output torque of the steering feel motor is designed as shown in (3.3).

$$T_{motor} = T_{alignment} + T_{assist} + T_{damping} + T_{\lim it} \tag{3.3}$$

3.2.3.1   Alignment Torque

The aligning torque plays a major role in steering feel. Although the aligning torque can be calculated by tire parameters, some information such as mechanical trail and pneumatic trail cannot be accurately obtained due to changeable driving conditions. Through the analysis of the steering mechanism, we found that the rack force is the main form of the aligning torque acting on the steering actuator.

Here, $T_{alignment}$ represents the alignment torque, which contains information about the road surface. It can be calculated by equation in (3.4).

$$T_{alignment} = -K_a(v) \cdot W(\alpha_f) \cdot F_{rack} \cdot r \qquad (3.4)$$

$$K_a(v) = \begin{cases} K_{normal}(v), & \theta_{sw} \cdot \dot{\theta}_{sw} > 0 \\ K_{return}(v), & \text{others} \end{cases} \qquad (3.5)$$

where $K_a(v)$ is a coefficient related to the velocity, r is the radius of the steering pinion. The calculation of the aligning torque can be transformed into the acquisition of the rack force F_{rack} and because of that the radius r is an inherent parameter. The estimation method to achieve rack force is discussed in the next section.

During the steering control of the vehicle, the control process of the steering wheel is divided into two stages, the normal process (when $\theta_{sw} \cdot \dot{\theta}_{sw} > 0$) and the return process. $K_a(v)$ takes different values in different stages so as to counteract part of the static friction of the steering wheel. α_f denotes front wheel slip angle, $W(\alpha_f)$ is the weighting function of the alignment torque related to front wheel slip angle. It is defined as (3.6), and the image is shown as Figure 3.3.

$$W(\alpha_f) = e^{-\frac{\alpha_f^2}{2 \cdot \sigma^2}} (1 - \kappa) + \kappa \qquad (3.6)$$

where, κ is a coefficient ranging from 0 to 1 that represents the lower bound of the weight function, and σ is the standard deviation of a Gaussian function. The front wheel slip angle can be obtained from a two-degree-of-freedom vehicle dynamic model as shown in Figure 3.4.

$$\alpha_f = \tan^{-1} \frac{v_{x,f}^t}{v_{y,f}^t} \qquad (3.7)$$

In the vehicle dynamic model, $v_{x,f}^t$, $v_{y,f}^t$ are the components of the front wheel velocity in the tire coordinate system. $v_{x,f}^o$, $v_{y,f}^o$ are the components of the front wheel velocity in the body coordinate system, and the coordinate transformation between the two coordinate systems is shown as (3.8) and (3.9). v_x, v_y are the longitudinal and lateral velocity of vehicle body, a is the distance from the center of mass of the vehicle to the front axle, $\dot{\omega}_r$ denotes the yaw rate of vehicle.

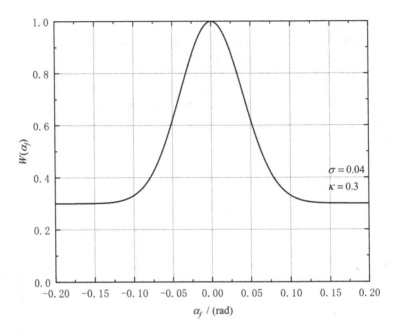

FIGURE 3.3   Image of the weighting function.
Source: Figure created by authors.

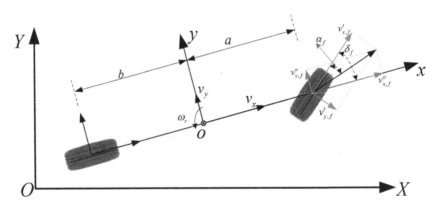

FIGURE 3.4   Schematic diagram of vehicle lateral dynamic model.
Source: Figure created by authors.

$$\begin{cases} v^t_{x,f} = v^o_{x,f} \cos \delta_f + v^o_{y,f} \sin \delta_f \\ v^t_{y,f} = v^o_{x,f} \sin \delta_f - v^o_{y,f} \cos \delta_f \end{cases} \qquad (3.8)$$

$$\begin{cases} v^{o}_{x,f} = v_{x} \\ v^{o}_{y,f} = v_{y} + a\dot{\omega}_{r} \end{cases} \tag{3.9}$$

3.2.3.2 Assisting Torque

During the process of steering control, the aligning torque is considered as a major component of the steering resistance load. In the conventional power assist steering system, the assisting torque can be generated by electric motor, which can reduce the steering resistance load on the handwheel. For the SBW system, the assisting torque is also necessary to be applied to the definition of steering feel, and allows the driver to have a comfortable steering control. Inspired by this, one method to design the assisting torque based on the steering handwheel angle. The calculation function of T_{assist} is expressed as

$$T_{assist} = \left(e^{\frac{-(k \cdot \theta_{sw})^2}{2\varepsilon^2}} (1-\rho) + \rho \right) \cdot G \tag{3.10}$$

where k is the coefficient for handwheel angle θ_{sw} and it is a constant, ε, ρ and G are the standard deviation, lower limit and gain of the assist function, respectively.

As shown in Figure 3.5, the function of assisting torque is maximum at zero handwheel angle and decreases with the increase of handwheel angle, which means that the assist effect increases as the angle increases. However, if the assist effect is extremely large when the steering angle is very large, the steering feel will be very light, which is not safe for the driver. The function of the lower limit is to prevent the assist effect from being infinite.

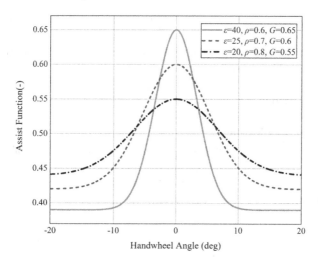

FIGURE 3.5 Graph of assisting torque function.

Source: Figure created by authors.

The assisting torque function in (3.10) is a kind of idealized numerical model. The dynamic disturbance of the steering system is not considered. In order to solve this issue, another method to define the assisting torque is considered by using the torque sensor on the steering column. The steering feel is mainly generated by the integration of system friction and assisting torque. T_{assist} is designed to counteract part of the system friction according to the torque of the steering column so that the remaining system friction torque is employed as the composition of steering feel.

$$T_{assist} = K_s(v) \cdot (e^{k_s \cdot T_s} - 1) \tag{3.11}$$

where, $K_s(v)$ decreases with increasing vehicle speed, T_s is the torque measured by the upper steering column sensor, k_s is a coefficient that determines the shape of the assisting curve.

3.2.3.3   Equivalent Friction Torque

The friction torque is an indispensable part of the steering feel in the conventional steering system. The dynamic friction model (e.g., LuGre friction model) can improve the emulation dynamics in both accuracy and stability. However, the design of dynamic friction model for mechanical system needs parameters' identification, which is conducted by a large number of experimental data analyses. The static friction model can meet the requirements and is often used to model the friction torque in steering feel design.

In this chapter, a hyperbolic tangent function model is employed to make the equivalent friction torque change smoothly, which is shown in Eqn. (3.12). The diagram of the equivalent friction torque is shown in Figure 3.6.

$$T_{friction} = \tau_{fs} \cdot \tanh(\eta \cdot \dot{\theta}_{sw}) \tag{3.12}$$

where τ_{fs} is the static friction constant of steering wheel system, which is determined through experimental parameter identification, $\dot{\theta}_{sw}$ is steering pinion angular velocity, and η is hyperbolic tangent coefficient which affects the slope of the equivalent friction torque curve.

3.2.3.4   Damping Torque

In order to prevent the steering wheel from overshooting, or instability due to excessive speed in the return process, $T_{damping}$ is designed to reduce the steering wheel reversing speed to ensure the safety of driving, which improves driving stability and maneuverability.

$$T_{damping} = -K_d(v) \cdot \tanh(k_d \cdot \dot{\theta}_{sw}) \tag{3.13}$$

where, $K_d(v)$ is a coefficient related to the velocity, k_d is a factor to reflect the influence of the angle rate.

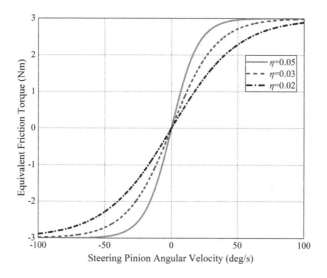

FIGURE 3.6 Graph of equivalent friction torque.

Source: Figure created by authors.

3.2.3.5 Limiting Torque

The steering feel motor should rapidly increase the resistant torque when the steering wheel angle reaches the threshold value. So that the limited working condition of the tire can be transmitted to the driver. The limited working strategy is designed as follows:

$$
T_{limit} = \begin{cases} k_l \cdot (\left|\theta_{sw}\right| - \theta_l), & \left|\theta_{sw}\right| > \theta_l > 0 \\ 0, & \left|\theta_{sw}\right| < \theta_l \end{cases}
\tag{3.14}
$$

where, k_l is the limiting torque adjustment factor, and θ_l is the absolute value of the limit position of the steering wheel angle.

3.2.4 Design of Rack Force Observer

Rack force is an important part of reconfigurable steering feel. Accurate evaluation of rack force is the basis and necessary condition to reflect the real road condition. In this subsection, an extended disturbance observer (EDO) is proposed to estimate rack force. The EDO method was originally designed based on the theory of the Romberg Observer. The EDO algorithm can reflect uncertainty in the system while the variables can be observed accurately and quickly. The previous research has shown that the extended disturbance observer has superior performance compared with the traditional disturbance observers, even in exceeding driving conditions [15].

In this section, the sum of the system friction of the steering actuator module τ_{fri} and rack force F_{rack} is considered as the generalized disturbance variable $d(t)$. It is assumed that $d(t)$ is a smooth variable with bounded derivation. To facilitate the calculation, Eqn. (3.2) is simplified to Eqn. (3.15).

$$\ddot{\theta}_m = -\lambda_1 \cdot \dot{\theta}_m + \lambda_2 \cdot \tau_m - \lambda_3 \cdot d \tag{3.15}$$

where, $\lambda_1 = \dfrac{B_m}{J_m}, \lambda_2 = \dfrac{i_{mc}}{J_m}, \lambda_3 = \dfrac{1}{J_m}$. State variables of the system are defined as $x_1 = \theta_m, x_2 = \dot{\theta}_m$. In this way, the state space equation of the second-order steering actuator is described by (3.16).

$$\begin{cases} \dot{x} = Ax + B\tau_m + Ed \\ y = Cx \end{cases} \tag{3.16}$$

The state matrix in (3.16) is expressed as $A = \begin{bmatrix} 0 & 1 \\ 0 & -\lambda_1 \end{bmatrix}; B = \begin{bmatrix} 0 \\ \lambda_2 \end{bmatrix}; C = \begin{bmatrix} 1 & 0 \end{bmatrix}; E = \begin{bmatrix} 0 \\ -\lambda_3 \end{bmatrix}$. θ_m is the angle of the steering column axis due to the equivalent conversion of the actuator, which can be regarded as the generalized rack angular displacement.

The original system is second order and expands to third order with the addition of the observed state for the generalized rack force as a disturbance. The extended disturbance observer design $L = \begin{bmatrix} u_1 & \mu_2 & \mu_3 \end{bmatrix}^T$ is shown in (3.17).

$$\begin{bmatrix} \dot{\hat{x}}_1 \\ \dot{\hat{x}}_2 \\ \dot{\hat{x}}_3 \end{bmatrix} = \underbrace{\begin{bmatrix} 0 & 1 & 0 \\ 0 & -\lambda_1 & -\lambda_3 \\ 0 & 0 & 0 \end{bmatrix}}_{\hat{A}} \begin{bmatrix} \hat{x}_1 \\ \hat{x}_2 \\ \hat{x}_3 \end{bmatrix} + \underbrace{\begin{bmatrix} 0 \\ \lambda_2 \\ 0 \end{bmatrix}}_{\hat{B}} \tau_m + \underbrace{\begin{bmatrix} \mu_1 \\ \mu_2 \\ \mu_3 \end{bmatrix}}_{L} (x_1 - \hat{x}_1)$$

$$\hat{d} = \underbrace{\begin{bmatrix} 0 & 0 & 1 \end{bmatrix}}_{\hat{C}} \begin{bmatrix} \hat{x}_1 \\ \hat{x}_2 \\ \hat{x}_3 \end{bmatrix} \tag{3.17}$$

where $\hat{A}, \hat{B}, \hat{C}$ represent the state matrix of the observer. $\hat{x}_1$ is the estimated value of the x_1. $x_1 - \hat{x}_1$ is equivalent to adding the output error to the system as a feedback quantity.

For computational purposes, the related parameters are defined as

$$\mu_1 = \frac{v_1}{\Delta}, \mu_2 = \frac{v_2}{\Delta^2}, \mu_3 = \frac{v_3}{\Delta^3} \tag{3.18}$$

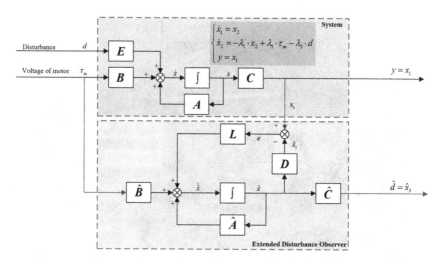

FIGURE 3.7 Schematic diagram of Extended Disturbance Observer.
Source: Figure created by authors.

where, v_1, v_2, v_3, Δ are the positive gain coefficients to be designed, which take values in the range of $[0,1]$. The system is a high-gain error feedback system, which can ensure the fast convergence of the observer. The rate of convergence of the system is related to Δ. The algorithm structure of the disturbance observer is illustrated in Figure 3.7, where $D = \begin{bmatrix} 1 & 0 & 0 \end{bmatrix}$.

3.3 EXPERIMENTS AND RESULTS ANALYSIS

3.3.1 VERIFICATION OF EDO ALGORITHM

In order to verify the feasibility and performance of the design method of reconfigurable steering feel and EDO algorithm, a hardware-in-the-loop (HIL) experimental platform was built (as shown in Figure 3.8), which includes power supply, steering wheel module, dSPACE, PC, and so forth. The experimental simulation environment is the automotive simulation models (ASM) in dSPACE. The dSPACE is a working platform for rapid control prototype, which has the advantages of high real-time, high reliability and good expandability. The software and hardware of the control system are based on the MATLAB/Simulink environment, and work as semi-physical simulation. The sampling time of dSPACE is set as 1ms and the frequency of steering wheel CAN signal is 10ms, so the rate transition module needs to be used in the simulink model to ensure the proper operation of the system.

To evaluate the performance of the rack force estimated by EDO algorithm, simulation experiment was done. The experimental scenario is the double lane

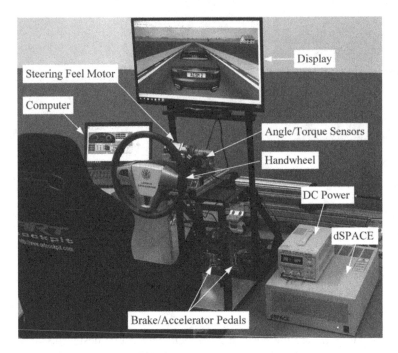

FIGURE 3.8   HIL experimental platform.
Source: Figure created by authors.

change condition and the speed of the vehicle is set at 75km/h. Figure 3.9 illustrates the performance of the EDO algorithm, in which the estimated values are compared with the theoretical reference values in the ASM model. Reference rack force can be followed accurately with low latency utilizing the EDO algorithm. The model parameters used in the experiments are shown in Table 3.1.

3.3.2   EVALUATION OF STEERING FORCE FEEDBACK

In the development of the conventional EPS system, a subjective evaluation method conducted by experienced drivers is often used to evaluate steering feel. In general, numerous iterative experiments should be carried out using the subjective evaluation method, which requires too much time and labor costs. The objective evaluation method, capable of reducing cost and speeding up development, is proposed in this chapter to evaluate steering feel. Based on the international standard for vehicle handling in ISO13674-1:2010, steering returnability, steering stiffness and steering on-center feel are employed as the evaluation metrics to objectively reflect the characteristics of steering force feedback. The definitions of these evaluation metrics are described in the following.

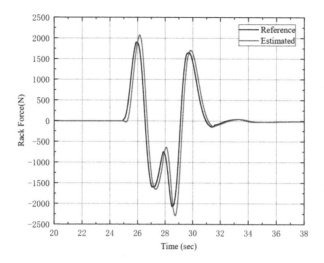

FIGURE 3.9   The performance of Extended Disturbance Observer.
Source: Figure created by authors.

TABLE 3.1
Parameters Used of EDO

Symbol	Description	Value
J_m	Equivalent moment of inertia	0.14
B_m	Equivalent damping coefficient	0.8
i_{mc}	Reduction ratio of actuator motor shaft to steering column shaft	21
r	radius of the pinion(mm)	7
v_1, v_2, v_3, Δ	Positive gain coefficients	$[2, 45, 10, 0.15]$

Source: Table created by authors.

Steering returnability (SR): As Figure 3.10(a) shows, when handwheel torque is zero, the value of the lateral acceleration is SR. Larger SR corresponds to a heavier steering feel.

Steering on-center feel (SF): As Figure 3.10(a) shows, when lateral acceleration is zero, the handwheel torque gradient value is SF. It represents the handling of a vehicle when the speed is high and lateral acceleration is low.

Steering stiffness (SS): As Figure 3.10(b) shows, when handwheel angle is between−10% and +10% of the range, the handwheel torque gradient value is SS. SS describes how clearly the driver perceives the handwheel center position. Larger SS corresponds to clearer center position.

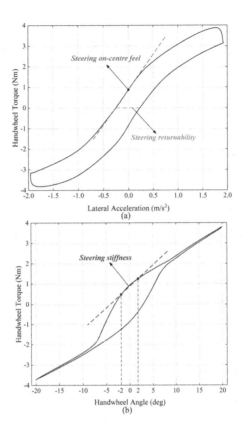

FIGURE 3.10   Objective evaluation metrics for steering feel.

Source: Figure created by authors.

Based on the above discussion of objective evaluation metrics for a traditional steering system, related objective indicators to evaluate steering feel on SBW can also be designed. Different drivers have different criteria for the evaluation of road feeling. The effect of steering feel on on-center handling performance for passenger cars is one of the typical methods for objective evaluation [16]. To carry out quantitative analysis for desired force feedback control of SBW, this section designed the following objective indicators to evaluate steering feel [17].

Angle Hysteresis (AH): is defined as the residual steering wheel angle when the steering wheel torque is zero and describes the hysteresis of the steering wheel angle with respect to the torque.

Steering Friction (SF): is defined as the value of residual torque when the steering wheel angle is zero and reflects the friction level of the steering system.

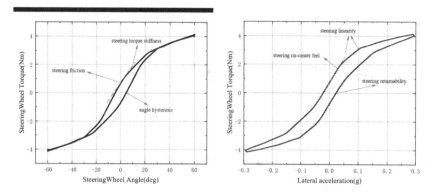

FIGURE 3.11 Steering wheel torque versus steering wheel angle and lateral acceleration. Source: Figure created by authors.

Steering Torque Stiffness (STS): denotes the torque gradient when the steering wheel angle is zero, which reflects the feeling of the steering wheel center position to the driver when driving at high speed.

Steering Returnability (SR): indicates the residual lateral acceleration when the steering wheel torque is zero. The smaller the value, the better the central area return performance.

Steering On-center Feel (SoF): is steering wheel torque gradient when the lateral acceleration is zero. This indicator directly reflects the handling characteristics of the vehicle. The larger the value, the clearer the steering feel.

Steering Linearity (SL): is defined as the ratio of torque gradient for lateral acceleration at 0.1g and steering on-center feel, describing the steering feel when the vehicle moves away from the central area in Figure 3.11.

3.3.3 STEERING FEEL IN DIFFERENT DRIVING CONDITIONS

3.3.3.1 Stationary Steering Test

At standstill, the experimental result of turning the steering wheel at 540 degrees left and right with a constant lower angular speed is shown in Figure 3.12. The black line refers to the steering wheel mechanical characteristics, while the red line presents the performance of the method of reconfigurable steering feel. The illustration indicates that the proposed approach can reduce part of friction appropriately.

3.3.3.2 Weave Steering Test

In order to verify the effectiveness of the proposed method of reconfigurable steering feel, and to study the influence of tuning parameters on the handling performance, on-center handling tests were conducted by sinusoidal steering control. The driver

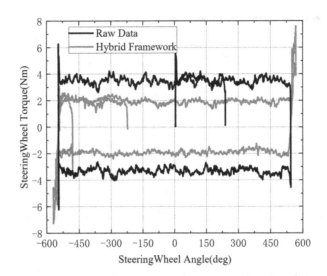

FIGURE 3.12   In-situ steering experiment.

Source: Figure created by authors.

performs the weave maneuver with a 0.2 Hz sinusoidal input according to the ISO standard (ISO13674-1:2010). During the weave maneuver test, the lateral acceleration does not exceed 2 m/s² and the vehicle speed is maintained at 60 km/h. The steering input of the driving control is shown Figure 3.13.

Figure 3.14 characterizes the effect of $k_a(v)$. The black line refers to the experimental results when the coefficients of the normal and return processes are equal. It can be seen that there is still a large residual dynamic and static friction at this time. The red line shows that the method of reconfigurable steering feel can effectively improve the effect of friction by taking different values at different processes.

Figures 3.15, 3.16, 3.17, 3.18 illustrate the effect of the main tuning parameters $k_s(v), k_s, k_d(v)$, k_d of the proposed method, respectively. The numerical result analyzed by objective evaluation indicators is listed in Table 3.2.

Table 3.3 summarizes the influence of the tuning parameters on steering feel according to the result of Table 3.2. In the tables, +refers to the effect of positive correlation, on the contrary, −indicates the effect of negative correlation and 0 indicates little significant correlation.

From the experimental data, it can be seen that the optimization of alignment torque can improve the return performance and angle hysteresis of steering feel. $K_s(v)$and $K_s$mainly affect steering torque stiffness, steering on-center feel and steering linearity. Steering feel can be improved by appropriately reducing the value of these two parameters. The two parameters $K_d(v)$and $K_d$affect the friction

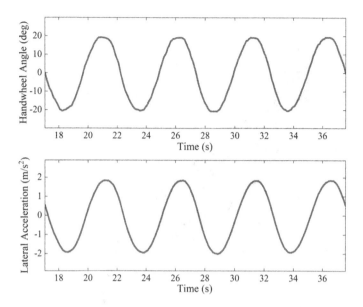

FIGURE 3.13 The steering input of the weave maneuver test.

Source: Figure created by authors.

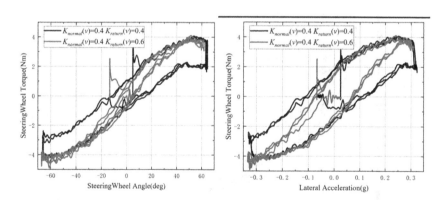

FIGURE 3.14 Steering wheel torque versus angle and lateral acceleration with different $k_a(v)$.

Source: Figure created by authors.

of the system. Increasing $K_d(v)$ can reduce angle hysteresis and steering friction, but at the same time, it will reduce the return performance of the steering wheel. On the contrary, increasing K_d can not only improve angle hysteresis and steering

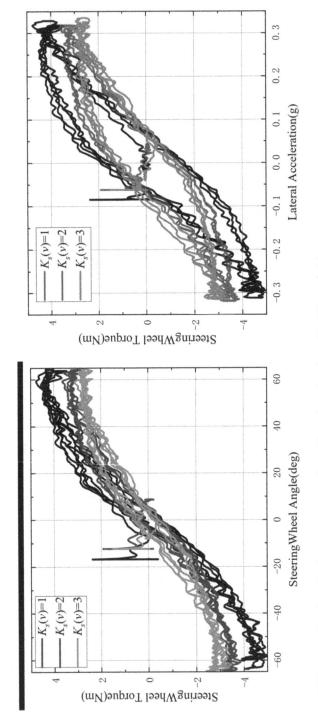

FIGURE 3.15 Steering wheel torque versus angle and lateral acceleration with different $k_s(v)$.

Source: Figure created by authors.

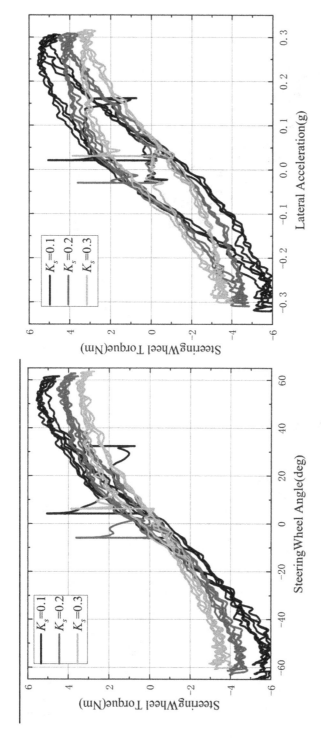

FIGURE 3.16 Steering wheel torque versus angle and lateral acceleration with different k_s.

Source: Figure created by authors.

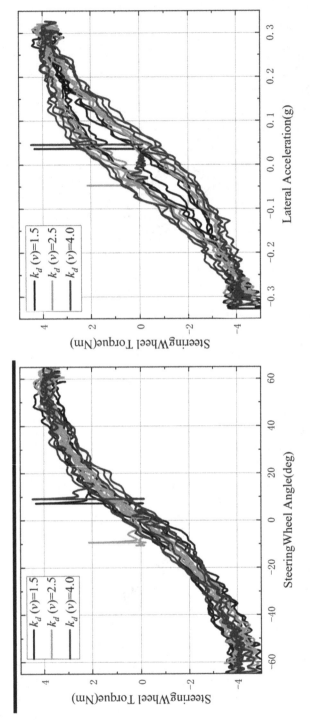

FIGURE 3.17 Steering wheel torque versus angle and lateral acceleration with different $k_d(v)$.

Source: Figure created by authors.

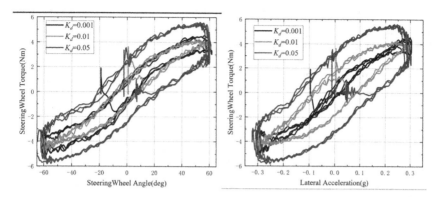

FIGURE 3.18 Steering wheel torque versus angle and lateral acceleration with different k_d.

Source: Figure created by authors.

TABLE 3.2
Steering Feel Performance at 60km/h

		AH	SF	STS	SoF	SR	SL
$k_s(v)$	1	4.5	0.496	0.150	22.088	0.070	0.696
	2	4.3	0.464	0.116	18.412	0.071	0.564
	3	4.2	0.448	0.080	15.501	0.076	0.515
k_s	0.1	4.2	0.680	0.164	27.279	0.063	0.612
	0.2	4.7	0.608	0.144	23.691	0.073	0.462
	0.3	4.6	0.496	0.122	20.784	0.072	0.337
$k_d(v)$	1.5	3.4	0.568	0.147	25.833	0.033	0.556
	2.5	2.1	0.208	0.129	23.355	0.056	0.554
	4.0	5.8	0.704	0.137	21.629	0.074	0.464
k_d	0.001	6.4	0.744	0.124	25.231	0.024	0.492
	0.01	8.8	1.056	0.116	20.111	0.085	0.606
	0.05	23.1	2.256	0.111	20.984	0.160	0.511

Source: Table created by authors.

TABLE 3.3
Influence of Steering Feel Parameters

	AH	SF	STS	SoF	SR	SL
$k_s(v)$	0	0	−	−	0	−
k_s	0	0	−	−	0	−
$k_d(v)$	−	−	0	0	+	+
k_d	+	+	0	0	+	0

Source: Table created by authors.

friction, but improve the return performance. In conclusion, the proposed hybrid framework to achieve the reconfigurable steering feel can be tuned to have realistic steering characteristics of vehicles, which provides feedback on dynamic load information in real time.

3.4 CONCLUSION

In this chapter, a hybrid framework is proposed to generate the reconfigurable steering feel for SBW, which mainly includes four parts: alignment torque, assisting torque, damping torque and limiting torque. The alignment torque is closely related to the rack force, and an extended disturbance observer is designed to estimate the rack force in real time. The alignment torque takes different coefficients in the normal and returns process. The friction torque assisting model is designed according to the torque of steering column to further optimize the smoothness of steering feel. The damping compensation prevents the steering wheel from returning too fast. The limiting torque prevents the steering wheel from exceeding its own cornering limit at a stationary operating condition. The analysis of influence of tuning parameters indicate that the proposed approach can realize the real-time feedback of loads from the tire–road interaction and have the reconfigurable steering characteristics of vehicles.

Based on the SBW vehicle dynamics coupled with the steering feel model, reconfigurable steering feel can be designed to provide the driver with a realistic steering feedback torque. The proposed parallel structure offers the possibility to introduce tuning factors that serve here as weights of different torque components, and could achieve a reconfigurable steering feel to meet the demands of individuation. The tuning parameters are associated with formal steering criteria or properties, and thus allowing a more rapid and intuitive tuning of the desired steering feel in order to meet the individual preferences of each driver. This kind of customizable characteristic is one of most important advantages for the SBW system.

REFERENCES

[1] Kirli, A., Arslan, M. S. (2016). Optimization of parameters in the hysteresis-based steering feel model for steer-by-wire systems. *IFAC Symposium on Control in Transportation Systems*.

[2] Fankem, S. and Muller, S. (2014). A new model to compute the desired steering torque for steer-by-wire vehicles and driving simulators. *Veh. Syst. Dyn.*, 52(sup1):251–271.

[3] Bhardwaj, A., Slavin, D., Walsh, J., Freudenberg, J. and Gillespie, R. B. (2021). Estimation and decomposition of rack force for driving on uneven roads. *Control Eng. Pract.*, 114:104876.

[4] Kant, N., Chitkara, R. and Pramod, P. (2021). *Modeling rack force for steering maneuvers in a stationary vehicle.* Technical report, SAE Technical Paper.

[5] Segawa, M., Kimura, S., Kada, T. and Nakano, S. (2004). A study on the relationship between vehicle behavior and steering wheel torque on steer by wire vehicles. *Veh. Syst. Dyn.*, 41: 202–211.

[6] Oh, S.-W., Yun, S.-C., Chae, H.-C., Jang, S.-H., Jang, J.-H., and Han, C.-S. (2003). *The development of an advanced control method for the steer-by-wire system to improve the vehicle maneuverability and stability.* SAE Technical Paper, SAE2003-01-0578.

[7] Salaani, M. K., Heydinger, G. J. and Grygier, P. A. (2004). *Closed loop steering system model for the national advanced driving simulator.* SAE Technical Paper, SAE 2004-01-1072.

[8] Mandhata, U., Wagner, J., Switzer, F., Dawson, D. M. and Summers, J. (2010). A customizable steer-by-wire interface for ground vehicles. *IFAC Proc. Vol.*, 43(7):656–661.

[9] Wang, J., Wang, H., Jiang, C., Cao, Z., Man, Z. and Chen, L. (2019). Steering feel design for steer-by-wire system on electric vehicles. In *2019 Chinese Control Conference (CCC)*, pages 533–538. IEEE.

[10] Asai, S., Kuroyanagi, H., Takeuchi, S., Takahashi, T. and Ogawa, S. (2004). *Development of a steer-by-wire system with force feedback using a disturbance observer.* SAE Technical Paper, 2004-01-1100, no. 724.

[11] Li, Y., Shim, T., Wang, D. and Offerle, T. (2017). Comparative study of rack force estimation for electric power assist steering system. In *Dynamic Systems and Control Conference*, 58295, page V003T33A005. American Society of Mechanical Engineers.

[12] Zhang, L., Meng, Q., Chen, H., Huang, Y., Liu, Y. and Guo, K. (2021). Kalman filter-based fusion estimation method of steering feedback torque for steer-by-wire systems. *Automot. Innov.*, 4:430–439.

[13] Balachandran, A. and Gerdes, J. C. (2014). Designing steering feel for steer-by-wire vehicles using objective measures. *IEEE/ASME Trans. Mechatron.*, 20(1):373–383.

[14] Yih, P. and Gerdes, J. C. (2005). Modification of vehicle handling characteristics via steer-by-wire. *IEEE Trans. Contr. Syst. Technol.*, 13(6):965–976.

[15] Wu, X., Zhang, M. and Xu, M. (2019). Active tracking control for steer-by-wire system with disturbance observer. *IEEE Trans. Veh. Technol.*, 68(6):5483–5493.

[16] Changfu, Z., Zhang, Z., Mai, L., Wang, C. and Wu, Z. (2013). *Study on objective evaluation index system of on-center handling for passenger car.* Technical report, SAE Technical Paper.

[17] Wang, C. Yan, L. Wu, X. and Kong, Z. (2023) *Design of Reconfigurable Steering Feel for Steer-by-Wire System Based on Dynamic Load Observation.* 7th CAA International Conference on Vehicular Control and Intelligence (CVCI), Changsha, China, pp. 1–7.

4 Shared Steering Control by Steer-by-Wire System

4.1 INTRODUCTION

With the development of automated driving technology, the intelligent vehicle has received growing attention in recent years for improving driving safety and reducing the human driver workload. However, there are still many challenges for the existing technologies to achieve a fully driverless vehicle in a complex environment [1]. Based on the Society of Automotive Engineers (SAE) standard, human-in-the-loop control is still a main technical approach for L0 ~ L3 intelligent vehicles before the L4 ~ L5 fully automated driving. The human–machine interaction technology is a kind of advanced driving assistance system (ADAS) that can be used to improve active safety and reduce driver workload [2]. A shared control concept by the combination of manual mode and automated mode offered a new solution to solving the control authority between the human driver and the machine systems [3].

During intelligent vehicle engineering practices, shared steering control is one of the important applications of the authority distribution for the driver assistance system. Shared control considers the human driver and the automated machine system as two intelligent agents. Based on the different work principles, shared control on the vehicle is usually implemented with single-agent mode or with the dual-agents mode. Single-agent mode means the human driver is considered as an independent driving input, and the machine system is used to enhance the driver's perception and control ability [4]. Human–machine interaction is a kind of master–slave relationship during shared control. The dual-agents mode takes the human driver and the machine system as two parallel control inputs. Based on the lane-changing situations, the control authority is switched between two intelligent agents. Two intelligent agents possess the control authority simultaneously, and generate coupling results according to a certain principle [5].

This chapter focuses on the modeling of human–machine cooperative steering control for an intelligent vehicle. There are two key issues during the design of shared steering control. The first is how to couple different inputs from two lane-changing participants. From previous literature, most of studies conducted

DOI: 10.1201/9781003481669-4

the integration of driver and machine steering input through an angle-coupling method. The control authority is distributed by a weight coefficient to couple inputs from human driver and machine system [6], [7], [8]. The shared-steering strategy is designed by mapping the path-tacking error onto the guiding steering angle. However, the angle coupling method usually generated unexpected steering results, thus leading to a confused sense for human handling reaction [9]. To build a feedback mechanism during human–machine interaction, this study has employed a torque-based coupling method to avoid possible violent steering guidance from the assistance system. The torque-based human–machine cooperative control allows the driver to remain in the control loop, and can achieve relative smooth authority distribution. This kind of approach is also studied as haptic shared control strategy in [10].

The second issue is how to deal with the conflict between human input and machine assistance. A variety of model-based human–machine cooperative method is designed for the shared steering controller, of which the model predictive control (MPC) [11], [12] and linear quadratic regulator (LQR) [13], [14] have been used to solve the human–machine interaction in trajectory tracking control. The model-based approaches with static numerical derivation are incapable of describing the dynamic characteristic of human driver reaction accurately. To describe and address conflicts between human and machine, game theory is an effective method to apply on such interactive decision-making system. The game-based approaches make the human and machine drive the vehicle in parallel, where the choice of steering action of each control agent depends on that of the other [15]. With regard to the research on the human–machine multi-agent system, since the advantages of game equilibrium strategy, this study investigated the shared steering control from a dynamic game theory approach. By considering the human–machine cooperative control as a continuous time-varying system, game-based shared steering control was first studied by X. Na [16], [17]. M. Flad employed motion primitives to fit the driver behavior and achieved cooperative lane-keeping control by differential games [18]. B. Ma proposed a non-cooperative game theoretic framework by illustrating the interaction on the steer-by-wire system considering human–machine goal consistency [19]. Shared steering control with a dynamic authority allocation strategy is investigated by a game-based model of predictive control presented in Song et al. [20]. An optimal vehicle stability control by coupling human driver and machine assistance is realized by using a preview game theory concept in Tamaddoni et al. [21]. By considering the course of the takeover and the actions from two participants, a framework enables obtaining the allocation of the control effort by calculating the Nash equilibrium between driver and machine assistance [22]. Most of these game-based approaches are designed by complete information hypothesis, where the game players can achieve full information from each other. The interaction of human and machine is conducted to follow a predefined principle during the conflict, and the control authority is assigned by an absolute trusting of the machine system. However, in the emergency lane-changing conditions with incomplete information, these kinds of game-based schemes usually generate machine-oriented shared results without enough attention to the driver's reaction. In order to address the authority distribution during lane-changing conditions, the intention of the human driver should be considered in the shared steering control. The attention of the human driver can be easily focused again after being reminded, but the decision of machine intelligence is limited to the performance of the sensory system. This study designs a steering torque cooperative framework for lane changing by using Nash equilibrium game theory. In the proposed framework, the shared control not only benefits from automated machine assistance but also makes full play of the human contribution. The human–machine interaction is modeled to enhance driving safety and simultaneously reduce the degree of conflict during shared steering control. Based on the experimental analysis in a hardware-in-the-loop (HIL) platform, the proposed strategy is verified by comparing the performance with the game algorithms from existing literatures.

Shared control structure is beneficial to the steering controller design of intelligence vehicles, and the human–machine goal consistency is a key prerequisite for shared control. However, the goal of consistency is usually given and cannot be changed, and a low goal consistency design of the steering controller directly deteriorates the vehicle performance in case of emergency – conditions that have not been sufficiently investigated. Based on the proposed shared control framework, the main tasks of the shared steering control by the steer-by-wire system are summarized as follows:

(1) An adaptive parameter adjustment scheme is designed to achieve dynamic Nash equilibrium during game control. The shared steering control is developed to obtain a reasonable game equilibrium by dynamically changing the authority distribution coefficient. A smooth resolution of human–machine conflict can be achieved by considering both driving comfort and safety.

(2) A torque-based shared steering control is employed to establish a flexible interaction mechanism in this study. Based on the conflict degree of steering torque, the machine system can reduce inappropriate assistance to follow the specific intention of the human driver. The proposed adaptive game approach can avoid single machine-oriented control results in which the human plays the dominative role in emergency driving conditions.

4.2 MODELLING OF THE SHARED STEERING CONTROL SYSTEM

4.2.1 SYSTEM ARCHITECTURE

The modeling of the human–machine system is designed by combining a single track vehicle dynamics model with a steer-by-wire (SBW) system. Compared with the traditional mechanical steering system, the SBW system removed the mechanical linkage between the steering column and steering actuator [23]. In this study, the human–machine shared steering control is investigated by using

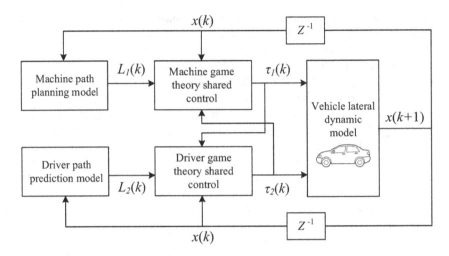

FIGURE 4.1   Architecture of the shared control model with game theory.

Source: Figure created by authors.

the decoupling characteristics of SBW systems. Based on the numerical models of both human driver and machine system, the target paths of two participants can be generated. The shared steering motion of the vehicle is a dynamic game process between the human driver and the machine system.

The system architecture of the shared steering control frame is shown in Figure 4.1. By considering the shared control as a linear-discretized system with time sequence k, here $L_1(k)$ and $L_2(k)$ denote the target trajectories of the machine system and human driver, respectively; $\tau_1(k)$ and $\tau_2(k)$ represent the steering torque generating from the machine and driver, respectively; $x(k)$ are the state parameters of the controlled vehicle dynamics, which are discretized as a time sequence. The interaction of human–machine shared control is coupled by steering torque from each other, and the vehicle is driven by the driver and machine simultaneously.

The human driver and machine system generate the optimal driving actions according to the respective target, which can be considered as a certain principle of game-based equilibrium. The game-based equilibrium usually generates compromise results to balance different requirements from two participants, which may not satisfy the traffic regulations. For example, the vehicle should be driven in the middle of the lane according to the traffic rules, but a neutral trajectory of game that results in making the vehicle run along the lane line between two traffic lanes will be unreasonable [5]. It is also difficult for the machine system to discover its inappropriate decisions under complex situations, such as driverless dilemma in [24].To solve the above issue, the game-based shared steering control can be designed as a human-oriented mode. This means the shared principle has the ability to return control authority to driver after the recovery of human attention. This chapter employs a torque-based shared steering control to establish a flexible human–machine interaction mechanism. When a conflict occurs, the machine will generate a reaction torque on the steering wheel to correct the steering action of the driver. The conflict can be resolved if the driver accepts the proposed action of the machine. But if the driver does not accept the assisted machine action, the driver will increase the steering torque to maintain the original operation. Here the accumulation of dot product between the steering torques from two participants is employed to recognize the driving attention of the human. More details of the control algorithm are given in the following section.

4.2.2 STEERING LATERAL DYNAMICS

In order to analyze the shared steering control, an integrated numerical model is considered by combining the vehicle dynamics with the steering system. To simplify the modeling of vehicle motion, a single-track dynamics model shown in Figure 4.2 is designed. With an assumption of constant longitudinal motion speed, the lateral dynamics equations of vehicle can be expressed as

$$m_c \left(\dot{v}_y + v_x \dot{\varphi} \right) = F_f + F_r \qquad (4.1a)$$

$$J_c \ddot{\varphi} = l_f F_f - l_r F_r \qquad (4.1b)$$

$$\dot{e}_\varphi = \dot{\varphi} - k \qquad (4.1c)$$

$$\dot{e}_y = v_x sine_\varphi + v_y cose_\varphi \qquad (4.1d)$$

$$\dot{s} = \frac{1}{1 - ke_y} \left(v_x cose_\varphi - v_y sine_\varphi \right) \qquad (4.1e)$$

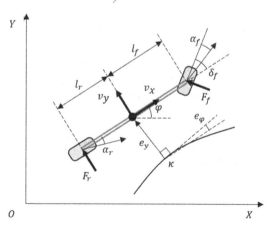

FIGURE 4.2   Vehicle single track dynamics model.
Source: Figure created by authors.

where m_c is the mass of the vehicle; J_c is the yaw moment of inertia of the vehicle; v_x and v_y are its longitudinal and lateral speeds; φ is the yaw angle of the vehicle relative to the road; l_f and l_r are the distances between the front / rear axle and the center of gravity of the vehicle, respectively; F_f and F_r are the lateral forces generated by the front and rear tires; e_φ is the included angle between vehicle heading direction and road center tangent; e_y is the lateral offset between vehicle and road centerline; k is the road curvature; $\dot{s}$ is the component speed along the road direction.

If the vehicle is driving on a road with little curvature ($k = 0$), the included angle between vehicle heading direction and road center tangent e_φ is assumed to keep in a relatively small range. Thus parts of the lateral dynamics equations of vehicle in Eqn. (4.1) can be rewritten as

$$\dot{e}_\varphi = \dot{\varphi}$$
$$\dot{e}_y = v_x e_\varphi + v_y \qquad (4.2)$$
$$\dot{s} = v_x - v_y e_\varphi$$

Based on the above assumption of small steering angle, the relationship between slip angle and lateral forces generated by the vehicle tires can be considered as approximate linear relation. Thus a linear tire model can be employed to approximately calculate the lateral force generated by the front and rear tires:

$$F_f = 2C_f \left(\frac{v_y + l_f \dot{\varphi}}{v_x} - \delta_f \right)$$
$$F_r = 2C_r \frac{v_y + l_r \dot{\varphi}}{v_x} \qquad (4.3)$$

where C_f and C_r represent the cornering stiffness of a single front tire or rear tire under its static load, and δ_f is front-wheel steering angle.

According to the previous study of steering system [23], the SBW system can be regarded as a rigid body. During the shared steering control, the modeling of handwheel module is considered as a one degree-of-freedom system. Based on Newton's law of motions the elements on the handwheel module are represented by its total moments of inertia and viscous friction coefficients. Based on this assumption, the steering dynamics function can be described as a second-order system, and the differential equation is given as

$$J_{SW} \ddot{\theta}_{SW} = \tau_1 + \tau_2 - \tau_{load} \qquad (4.4)$$

where J_{SW} is the equivalent moment of inertia of the steering system with respect to the steering wheel, θ_{SW} is the steering angle of handwheel, τ_1 is the assist torque with respect to the steering wheel provided by the machine system, τ_2 is the steering torque of the driver. τ_{load} is the equivalent steering load, which is composed of steering systems equivalent viscous friction torque and self-aligning torque. The self-aligning torque is calculated with respect to the steering wheel pneumatic trail in [25], the calculation of τ_{load} can be expressed as

$$\tau_{load} = B_{SW}\,\ddot{\theta}_{SW} + \frac{2C_f d_w}{N_S}\left(\delta_f - \frac{v_y + d_p\dot{\varphi}}{v_x}\right) \tag{4.5}$$

where d_w is the width of the front wheel, d_p is the pneumatic trail of front wheel, and B_{SW} is the equivalent viscous friction damping coefficient of the steering system.

The steering transmission from the steering wheel angle θ_{SW} to the front wheel steering angle δ_f is assumed as a linear relation. If the transfer ratio is defined as N_s, the relation can be described as $\theta_{SW} = \delta_f N_s$. Substituting Eqn. (4.4) and (4.5) into Eqn. (4.1), and denoting the lateral deviation of the vehicle from its starting point as Y, the system state equation of vehicle dynamic is expressed as

$$\begin{cases} \dot{x} = Ax + B\left(\tau_1 + \tau_2\right) \\ y = Cx \end{cases} \tag{4.6}$$

where

$$x = \begin{bmatrix} \theta_{SW} & \dot{s}_{SW} & v_y & \dot{\varphi} & Y & \varphi \end{bmatrix}^T$$

$$A = \begin{bmatrix} 0 & 1 & 0 & 0 & 0 & 0 \\ -\dfrac{c_1}{N_S} & -\dfrac{B_{SW}}{J_{SW}} & \dfrac{2C_1}{v_x} & \dfrac{2C_1 d_p}{v_x} & 0 & 0 \\ -\dfrac{2C_f}{m_c N_s} & 0 & C_2 & C_3 - v_x & 0 & 0 \\ -\dfrac{2l_f C_f}{I_c N_s} & 0 & C_3 & C_4 & 0 & 0 \\ 0 & 0 & 0 & 0 & 0 & v_x \\ 0 & 0 & 0 & 1 & 0 & 0 \end{bmatrix}$$

$$B = \begin{bmatrix} 0 & \dfrac{1}{J_{SW}} & 0 & 0 & 0 & 0 \end{bmatrix}^T$$

$$C = \begin{bmatrix} 0 & 0 & 0 & 0 & 1 & 0 \\ 0 & 0 & 0 & 0 & 0 & 1 \end{bmatrix}$$

$$C_1 = \frac{2C_f d_w}{N_s J_{SW}}$$

$$C_2 = \frac{2C_f + C_r}{m_c v_x}$$

$$C_3 = \frac{2l_f C_f - 2l_r C_r}{m_c v_x}$$

$$C_4 = \frac{2l_f^2 C_f + 2l_r^2 C_r}{m_c v_x}$$

In order to be used in the continuous control model, the above system state equation needs to be converted to the discrete form with the sampling time T_s. If k is defined as the discrete time instance, the equation can be rewritten as

$$\begin{cases} x(k+1) = A_d x(k) + B_d \left(\tau_1 + \tau_2 \right) \\ \qquad\qquad y(k) = Cx(k) \end{cases} \tag{4.7}$$

where

$$A_d = e^{A T s}$$

$$B_d = B \int_0^{T_s} e^{A t} dt$$

In this study, the related physical parameters of above vehicle dynamics model are shown in Table 4.1.

4.2.3 MACHINE SYSTEM MODELLING

During the shared control of the vehicle, the control model of machine intelligence should first be designed. The target vehicle should be controlled to track a safe trajectory generating from path planning algorithms. The parameters of vehicle dynamic model are shown in Figure 4.2. Based on the assumption of vehicle movement with constant longitudinal velocity, the kinematic model of vehicle to describe the motion trajectory is expressed as:

TABLE 4.1
Physical Parameters of the Vehicle Dynamics

Symbol	Description	Value
m_c	Vehicle mass	$1406kg$
J_c	Vehicle yaw moment of inertia	$1802kg{\cdot}m^2$
v_x	Vehicle cruising speed	$20m/s$
l_f	Distance from vehicle center of gravity to front axle	$1.016m$
l_r	Distance from vehicle center of gravity to rear axle	$1.562m$
C_f	Cornering stiffness of one single front wheel	$70000N/rad$
C_r	Cornering stiffness of one single rear wheel	$50000N/rad$
J_{sw}	Moment of inertia of steering system	$0.1kg{\cdot}m^2$
B_{sw}	Steering system linear damping coefficient	$0.8Nm{\cdot}s/rad$
N_s	Steering system angle reduction ratio	16
d_w	Width of front wheel	$0.2m$
d_p	Front wheel pneumatic trail	$0.008m$

Source: Table created by authors.

$$\dot{v}_y = a_y$$

$$\dot{v}_x = 0$$

$$\dot{\varphi} = \frac{a_y}{v_x} \quad (4.8)$$

$$\dot{Y} = v_x sin\varphi + v_y cos\varphi$$

$$\dot{X} = v_x cos\varphi - v_y sin\varphi$$

where a_y is the lateral acceleration of the vehicle, X and Y are the vehicle's global longitudinal displacement and lateral displacement relative to its starting point.

Under the premise that φ is sufficiently small, Eqn. (4.8) can be linear-discretized with time step T_s as

$$v_y(k+1) - v_y(k) = T_s a_y(k)$$

$$v_x(k+1) - v_x(k) = 0$$

$$\varphi(k+1) - \varphi(k) = \frac{T_s a_y(k)}{v_x(k)} \quad (4.9)$$

$$Y(k+1) - Y(k) = T_s\left[v_x(k)\varphi(k) + v_y(k)\right]$$

$$X(k+1) - X(k) = T_s\left[v_x(k) - v_y(k)\varphi(k)\right]$$

where k is the current discrete time instance. Denoting the machines predicted time domain length as N_p, the planned path for machine system can be expressed as

$$L_1(k) = \begin{bmatrix} y(k) & y(k+1) & \dots & y(k+N_p-1) \end{bmatrix}^T \qquad (4.10)$$

where

$$y(k) = \begin{bmatrix} Y(k) & \varphi(k) \end{bmatrix}^T$$

The planning algorithm of obstacle avoidance function is realized by a time-variant safety margin method from [26]. The specific procedure starts by introducing a cost function $z_o(j)$ to describe the safety margin between the vehicle and obstacle:

$$z_o(j) = \frac{1}{\left(X(j) - X_o(j)\right)^2 \left(Y(j) - Y_o(j)\right)^2} \qquad (4.11)$$

where $j = k, \dots, k+N_p-1$, $X_o(j)$, and $Y_o(j)$ are the estimated longitudinal and lateral positions of the obstacle at time sequence j, which are given by:

$$\begin{cases} X_o(j) = X_o(k) + v_{xo}(k)(j-k)T_s \\ Y_o(j) = Y_o(k) + v_{yo}(k)(j-k)T_s \end{cases} \qquad (4.12)$$

where $X_o(k)$ and $Y_o(k)$ are measured positions of the obstacle at current time instance k, $v_{xo}(k)$ and $v_{yo}(k)$ are corresponding speeds in longitudinal direction and lateral direction respectively.

The path planning algorithms can be designed as an optimization problem for vehicle lateral acceleration. The solution of control sequence is defined as $\begin{bmatrix} a_y(k) \dots a_y(k+N_p-1) \end{bmatrix}^T$. The optimal control sequence of vehicle are calculated by a quadratic objective function over a prediction horizon of N_p sampling intervals, which is defined as

$$J(k) = \sum_{j=k}^{k+N_p-1} \frac{1}{2} [\varphi(j)^T W_\varphi \varphi(j) + z_o(j)^T W_o z_o(j) + a_y(j)^T W_a a_y(j)] \qquad (4.13)$$

where W_φ, W_o, and W_a are scalar weighting coefficients related to heading angle, distance to obstacle and lateral acceleration, respectively.

Based on the solution of control sequence by the Eqn. (4.9) and Eqn. (4.13), the planned path from machine intelligence can be generated for the obstacle avoidance.

4.2.4 MODELLING OF DRIVER PATH-FOLLOWING CONTROL

To investigate the human–machine shared steering control of the vehicle, a numerical driver model is needed to generate input from the human side. Here, a two-point preview driver model is employed as a reasonable approximation of the driver's steering behavior [27], which is shown in Figure 4.3. The model defines two preview points on the road along the driving direction of the vehicle. The near point is located on the center line of the lane, with a distance D_n ahead. The far point is also located on the center line of the lane with a distance D_f ahead when the road is straight, and it is defined as the point of contact on the tangent line from the vehicle to the inner edge of the lane when the road is curved. The angle between vehicle's velocity vector and the lines connecting the vehicle to two preview points are denoted as near angle θ_n and far angle θ_f respectively. Here the included angle is approximately equal to the sinusoidal values, the calculation can be written as

$$\begin{cases} \theta_n = \dfrac{Y_n - Y}{D_n} - \varphi \\ \\ \theta_f = \dfrac{Y_f - Y}{D_f} - \varphi \end{cases} \tag{4.14}$$

where Y_n and Y_f are the lateral position of near preview point and far preview point of the driver. By using a proportional differential preview control method, the steering wheel angular velocity expected by the driver is calculated by

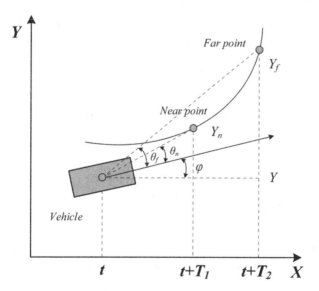

FIGURE 4.3 Two points preview control of driver model.
Source: Figure created by authors.

$$\dot{\theta}_{SW} = \frac{K_p \left(\theta_n + \theta_f \right)}{2} + K_f \dot{\theta}_f + K_n \dot{\theta}_n \qquad (4.15)$$

where K_f and K_n are the gain coefficients corresponding to the changing rate of the far angle and near angle, K_p is the coefficients value from heading angle to steering angle.

Note that the $\dot{\theta}_{SW}$ obtained in this way is a single-step control value of driver model, a continuous reference path expected by the driver can be calculated with the above vehicle lateral dynamic. Therefore, in the prediction time domain corresponding to current time step k, the value of $\dot{\theta}_{SW}$ calculated at time step $k+j$ is used to predict the vehicles new state at time $k+j+1$. The sequence of driver input is updated by the rolling prediction of vehicle state corresponding to the increased time step. The calculation of vehicle state here is similar to the system state equation in Eqn. (4.6), and is expressed as:

$$\begin{cases} \dot{x}_{drv} = A_{drv} x_{drv} + B_{drv} \dot{\theta}_{SW} \\ y_{drv} = C_{drv} x_{drv} \end{cases} \qquad (4.16)$$

where

$$x = \begin{bmatrix} \theta_{SW} & v_y & \dot{\varphi} & Y & \varphi \end{bmatrix}^T$$

$$A_{drv} = \begin{bmatrix} 0 & 0 & 0 & 0 & 0 \\ -\dfrac{2C_f}{m_c N_s} & C_2 & C_3 - v_x & 0 & 0 \\ -\dfrac{2l_f C_f}{I_c N_s} & C_3 & C_4 & 0 & 0 \\ 0 & 0 & 0 & 0 & v_x \\ 0 & 0 & 1 & 0 & 0 \end{bmatrix}$$

$$B_{drv} = \begin{bmatrix} 0 & 0 & 0 & 0 & 0 \end{bmatrix}^T$$

$$C_{drv} = \begin{bmatrix} 0 & 0 & 0 & 1 & 0 \\ 0 & 0 & 0 & 0 & 1 \end{bmatrix}$$

Denoting the length of drivers prediction time domain as N_p, the predict sequence of the driver reference path at time instance k is expressed as

$$L_2(k) = \begin{bmatrix} y_{drv}(k) & y_{drv}(k+1) & \cdots & y_{drv}(k+N_p-1) \end{bmatrix}^T \qquad (4.17)$$

TABLE 4.2
Parameters of Adaptive Game Control Strategy

Symbol	Description	Value
N_p	Control prediction time domain length	40
$a_{y,max}$	Upper limit of lateral acceleration value	$2m/s^2$
W_φ	Heading angle weight coefficient	1000
W_o	Obstacle distance weight coefficient	500
W_a	Lateral acceleration weight coefficient	2
D_n	Near point distance	$6.2m$
D_f	Far point distance	$20m$
K_f	Far angle differential gain	20
K_n	Near angle differential gain	20
K_p	Near angle proportional gain	20
q_Y	Weight coefficient of lateral displacement error	10
q_φ	Weight coefficient of yaw angle tracking error	0.01
R_0	Baseline value of weight coefficient for the machine steering torque	1
α	Base value of exponential function	$10^{0.1}$

Source: Table created by authors.

The related parameters cited above are shown in Table 4.2. From the above driver path modeling, the expected sequence of driver steering angle θ_{sw} can be calculated. Since the following human–machine shared control is torque-based model, the driver steering angle can be converted into the steering torque of driver by using steering dynamics function in Eqn. (4.5).

4.3 DESIGN OF GAME-BASED SHARED CONTROL ALGORITHM

During the human–machine interactive steering control, the motion of vehicle is commanded at the same time from two intelligence agents. The specific shared control process between the machine system and the human driver can be viewed as a series of non-cooperative Nash games [16]. In this type of game, the steering torques from machine intelligence and human driver are coupled to generate final steering action. The machine system and the human driver formulate their own strategies independently, but are aware of each other's actions and their possible impact on vehicle movement. When the prediction time domains of the machine and the driver are defined to have the same length N_p, the cost functions for the torque inputs from both participants at time instance k are given as

$$J_1\left(\tau_1+\tau_2\right)= \sum_{j=k}^{k+N_p-1} \left\|y(j)-y_1(j)\right\|_{Q_1}^2 + \sum_{j=k}^{k+N_p-1} R_1\tau_1(j)^2$$

$$J_2\left(\tau_1+\tau_2\right)= \sum_{j=k}^{k+N_p-1} \left\|y(j)-y_2(j)\right\|_{Q_2}^2 + \sum_{j=k}^{k+N_p-1} R_2\tau_2(j)^2$$

$$(4.18)$$

where $\tau_1(j)$ and $\tau_2(j)$ are the torque input sequences from machine system and human driver, $y_1(j)$ and $y_2(j)$ are the target planned path sequences from machine system and human driver, $y(j)$ is the synthetic path sequence by human–machine interaction. $Q_i = diag\left(q_y,q_\varphi\right)$ is the weight coefficient matrix of lateral displacement and yaw angle tracking error, which is corresponding to the tracking performance of vehicle. R_i is the weight coefficient of the amount of steering torque input which is corresponding to the magnitude of control effort.

If τ_i^* is defined as the optimized steering torque allocation strategy under Nash equilibrium. The target planned path $y_i(j)$ can be obtained from Eqn. (4.10) and (4.17). Here a dynamic programming inverse order method is applied for the solution of the τ_i^*. Before using the linear quadratic programming algorithm, Eqn. (4.18) needs to be converted to its standard format:

$$\tau_i^* = \arg\min_{\tau_i}[\sum_{j=k}^{k+N_p-1} \left\|\tilde{x}(j)\right\|_{\tilde{Q}_i}^2 + \sum_{j=k}^{k+N_p-1} R_i\tau_i(j)^2],i=1,2 \qquad (4.19)$$

where

$$\tilde{x}(j)=\left[y(j) \quad L_i(k)\right]^T$$

$$\tilde{Q}_i = \begin{bmatrix} K_i^T Q_i K_i & -diag\left(y_i^T Q_i\right) \\ -diag\left(Q_i^T y_i\right) & 0 \end{bmatrix}$$

$$K_1 = \left[C \quad -I_2 \quad 0_{2\times(2N_p-1)}\right]^T$$

$$K_2 = \left[C \quad 0_{2\times N_p} \quad -I_2 \quad 0_{2\times(N_p-1)}\right]^T$$

$$(4.20)$$

According to Hamilton formalism, the optimal torque input of the participants under Nash equilibrium has the following form

$$\tau_i^* = -R_i^{-1}\tilde{B}_i^T P_i \tilde{x} \qquad (4.21)$$

We refer to $-R_i^{-1}\tilde{B}_i^T P_i$ as the torque feedback matrix under Nash equilibrium, where $\tilde{B}_i = \left[B_i \quad 0\right]^T$. Matrix p_i is obtained by solving a series of discrete Riccati equations:

$$P_i = P_i^0$$

$$P_i^{k-1} = P_i^k + T_s \left[\left(\tilde{A}^k - S_{3-i}^k P_{3-i}^k \right)^T P_i^k + P_i^k \left(\tilde{A}^k - S_{3-i}^k P_{3-i}^k \right)^T - P_i^k S_i^k P_i^k + \tilde{Q}_i^k \right]$$

$$S_i^k = \widetilde{B}_i R_i^{-1} \tilde{B}_i^T$$

$$P_i^{N_p} = 0$$

(4.22)

During the solution of above algorithm, the iterative calculation of $P_i^{N_p}$ is started with initial value as zero. Based on the value of P_i^k, P_i^{k-1} can be derived by the forward iterative calculation. By using rolling optimization approach, the results of τ_i^* can be recalculated in each computing cycle.

The game-based functions in Eqn. (4.19) describe the torque interactive between machine system and human driver. It can be observed from the cost function that the larger Q_i / R_i is, the more attention the participant pays to the path tracking accuracy. Since R_1 and R_2 are the weight coefficients of the amount of steering torque input corresponding to the machine intelligence and human intelligence, the allocation of shared steering control can be conducted by the adjusting of related weight coefficient.

During normal driving scenarios, the human driver plays the main role in the control of the vehicle, and the assistance from the machine system is used to decrease the driver's workload and improve the driving experience. During the emergency scenarios, the machine intelligence compensates the human drivers control input to reduce the unsafe driving actions. As a result, the design principle of shared steering control is based on the safety of the coupled reference path. However, the game-based control strategies from two participants usually generate synthetic paths to meet their respective requirements. This kind of compromise reference path should be checked to find a collision-free trajectory. To simplify the algorithm, here the machine assistance system is supposed to have information of the driver's reference path. By using the vehicle system state Eqn. from the Eqn. (4.7), the discrete form of synthetic path by human–machine interaction can be expressed by

$$X_{pr}(i) = X_0 + kv_x T_s, 1 \le i \le N_p,$$

$$x_{pr}(k+i) = A_d x_{pr}(k+i-1) + B_d \left[\tau_1^*(k+i) + \tau_2^*(k+i) \right],$$

(4.23)

$$Y_{pr}(i) = Cx_{pr}(k+i).$$

where X_{pr} and Y_{pr} are the longitudinal and lateral coordinate sequences of the predicted synthetic path respectively, whose lengths are both N_p; $x_{pr}(k+i)$ is the vehicle state prediction vector at time step $k+i$; τ_1^* and τ_2^* are the optimal torque inputs from the machine and driver respectively, which is calculated by Eqn. (4.19).

Based on the path coordinate sequence (X_{pr}, Y_{pr}), the collision-free trajectory can be checked by the analysis of distance between each coordinate point and the obstacles. If the synthetic path has potential safety hazards, the control weight of machine assistance system will be increased to correct unsafe behavior. Otherwise, the assistance from the machine system should be reduced to minimum to avoid intervention of the human driver. The weight coefficient R_i in Eqn. (4.19) can be employed to allocate steering control right between human driver and machine system. Here the definition of R_2 is set as a constant value, the rule of adaptive adjustment of R_1 can be formulated by establishing a value pool as $R_{1,lst}$. The $n-th$ element in the array is defined as

$$R_{1,lst}\left(n\right) = R_0\alpha^{\beta-n}, n = 0,1,2\ldots,\tag{4.24}$$

where R_0 is the baseline value of R_1, which is set to provide an appropriate balance between path tracking accuracy and control effort magnitude. β is the exponent upper bound for generator. The related parameters of game control strategy are shown in Table 4.2. To obtain a collision-free trajectory from the synthetic paths, the weight coefficient R_1 is selected from the value pool in sequence. The initial value R_1 with $n = 0$ means a human-controlled model where the machine system has minimum control intervention. As the sequence n rises, the value of R_1 decreases to achieve a larger authority distribution of machine system. For each calculation cycle, the safety of synthetic path with selected R_1 is checked by the collision cost function in Eqn. (4.11). If the game result has potential collision risk, it will in–crease machine intervention by reselecting a smaller R_1 in sequence until the synthetic path is safe. The value of R_1 is selected in ascending order to not only find a minimum machine assistance but also have a collision-free shared path.

From the above discussion, the shared steering control of the vehicle is still a machine-oriented mode for the emergency treatment. Since the control strategies of machine intelligence can not cover all possible driving scenarios, a human-oriented mode is necessary to be considered to return final control right to driver. The real intention of human driver is used to correct the authority distribution. The lane change intention can be identified by the feature vectors of vehicle state of longitudinal and lateral dynamics [28]. To describe the confliction between the two participants, the dot product of the torques from human driver and machine system can be accumulated to show the degree of conflict. Here the adaptive adjustment of R1 is redesigned by a range limitation of exponent upper bound. The calculation of Eqn. (4.24) is rewritten by

$$R_{1,lst}\left(n\right) = R_0\alpha^{\beta^*-n}, n = 0,1,2\ldots,$$
$$\beta^* = \beta_0\left[\sum_{j=k}^{N_p+k}\frac{1}{2}[\tau_1\tau_2]\right]\tag{4.25}$$

where β_0 is the gain coefficient of integral accumulation. An increased cumulative quantity of the steering torques represents a high intervention from human driver, which is also used to recognize the degree of driving attention. The value of R_1 is designed based on the conflict degree of steering torque, the machine system can reduce unnecessary assistance to follow the driving intention of human driver. This kind of adaptive game control strategy can achieve a dynamic authority distribution by considering both of driving comfort and safety.

4.4   EXPERIMENT AND ANALYSIS

In order to verify the performance of game-based shared steering control, hardware-in-the-loop (HIL) experiments have been conducted on a self-developed driving simulator. The structure of the experimental platform for steer-by-wire system is shown in Figure 4.4. The steering assisted motor on the handwheel module is used to generate control torque from machine intelligent system. The handwheel angle and torque signals are collected by installed sensors. The real-time control is conducted by a rapid control prototype dSPACE PX20. The game-based algorithmic model is built under MATLAB/Simulink environment. The sampling rate of the system is selected as $10\,ms$. CAN bus communication is applied between the handwheel module and the real-time controller. To simulate a human–machine interaction progress, traffic scenarios with vehicle dynamic model are constructed in the automotive simulation environment.

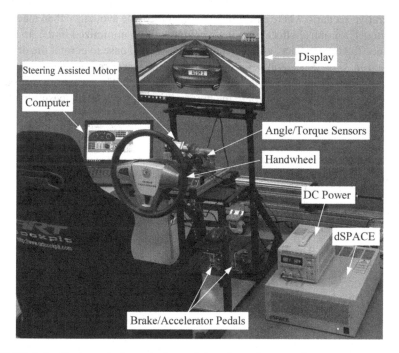

FIGURE 4.4   Hardware-in-loop platform of shared steering control.

Source: Figure created by authors.

Based on the degree of consistency of control inputs from human driver and machine system, different traffic scenarios are designed to show the efficiency of proposed game-based shared steering algorithm. The test is conducted on the previously mentioned HIL experimental platform. The performance of shared control with human–machine consistent goal and conflict goal are analyzed in an obstacle-avoidance scenario. The human participant conducted the experiments after becoming completely familiar with the HIL platform. For each of the following experimental scenarios, the performance of the human–machine shared steering control with adaptive game parameters is analyzed by comparing the control results with the conventional game-based assistance with fixed parameters in [18].

In this chapter, the steering behavior of the human driver is performed under the limit-handling condition to reduce the influence of psychological factors. Therefore, the drivers' target paths are predefined to validate the proposed control model. The paths to be tracked during the collision avoidance are shown in Figure 4.5. The road to traverse is configured as a straight lane with the controlled vehicle starting from the left side, and a single static obstacle is located in front of the vehicle. Based on the state of driver attention, three kinds of scenarios with different driver target paths were analyzed to show the effect of shared steering control [29]. Case 1 is a cooperative scenario where the driver makes an earlier obstacle avoidance than the planned results of the machine. Case 2 is a cooperative scenario where driver makes a later obstacle avoidance than machines planning. The control effect from the machine will be minimized in the first case to reduce the interference, and increased in the second case to correct the driver's later action. Case 3 is a conflict scenario where the driver resists against obstacle avoidance by maintaining a forward-driving behavior. This scenario is used to verify the return of final control to the human driver if the human doesn't trust the machine's decision.

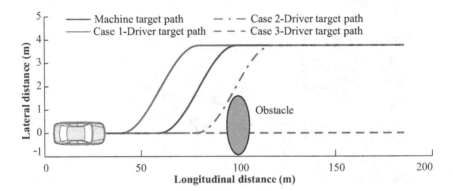

FIGURE 4.5   Target paths of machine system and human driver in different scenarios.
Source: Figure created by authors.

4.4.1 CASE 1: SCENARIO WITH EARLY OBSTACLE AVOIDANCE

In this scenario, the vehicle is driving with constant longitudinal velocity from the beginning of the experiment. From the general automobile safety standards, the safety distance is not less than 30m if the vehicle is driving with speed of 40km/ h [14]. Here the velocity of a simulated vehicle is set as 40km/h. The driver is assumed to notice the danger at 40m longitudinal position before the obstacle, and performs an early lane change movement to avoid the obstacles in a timely way. Meanwhile, the machine intelligent system is assumed to take avoidance action at 60m longitudinal position before the obstacle. The proposed adaptive game-based shared steering control is compared with a previous shared control approach with fixed game parameters. The simulation result is shown in Figure 4.6.

Since the target path of the driver is different from that of machine planning, it can be found from Figure 4.6 that the human driver took a timely action to

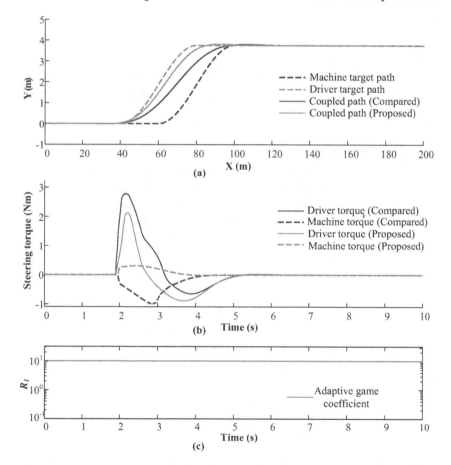

FIGURE 4.6 Experimental result when driver makes a timely response to avoid obstacle.
Source: Figure created by authors.

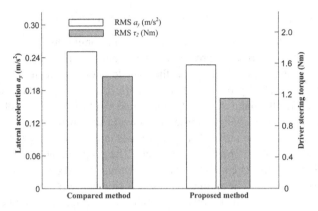

FIGURE 4.7 Performance analysis of different control approaches.

Source: Figure created by authors.

avoid the obstacle in advance. The synthetic path by the compared method is a compromise result, where the vehicle runs along the neutral trajectory between two target paths. Since the prediction of the human driver is a safe action, the synthetic path by the proposed method can achieve a human-oriented result to reduce the intervention from the machine system. The steering torque shows the conflict of driver and machine system. The proposed method has a smaller steering torque, which is benefit from the adaptive game-based shared control.

From Figure 4.7, the root mean square (RMS) of lateral acceleration and driver steering torque are selected as indexes to evaluate the performance of the proposed shared control method. Lateral acceleration and driver steering torque can be used to show the smoothness and comfort during the avoidance motion. The results of two indexes can be used to evaluate the degree of competition during the shared control. It can be found that the proposed method can reduce the conflict of steering torques from two participants, which brings an improved driving experience.

4.4.2 CASE 2: SCENARIO WITH LATE OBSTACLE AVOIDANCE

The design of this scenario is similar with that of Case 1. To avoid the static obstacle located in front of the driving vehicle, the machine system is still assumed to take avoidance action at 60 m longitudinal position before the obstacle. The difference is that the driver is assumed to take a later action due to the inattention of human. The drivers target path of lane change is started at 80m longitudinal position before the obstacle, where is difficult to conduct available behavior to avoid obstacle. Here the shared control model will be employed to correct the unsafe behavior of the driver. The simulation result is shown in Figure 4.8.

From the experimental results in Figure 4.8, it can be found that synthetic paths to avoid obstacle are achieved with the assistance of shared control between

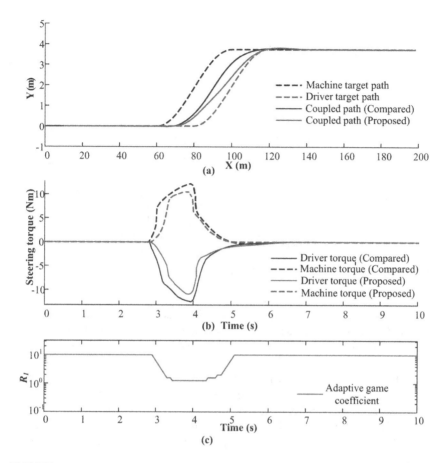

FIGURE 4.8 Experimental result when driver make a late response to avoid obstacle.
Source: Figure created by authors.

human driver and machine system. The authority distribution parameter of game controller R1 is adjusted to balance the requirements from two participants. Based on the adaptive game control strategy in Eqn. (4.25), the value of R1 is reduced to increase the authority distribution of machine assistance. Since the compared method is conducted with fixed game parameter R1, the synthetic path is a compromise result with more serious torque conflict. From the evaluation results of lateral acceleration and driver steering torque in Figure 4.9, the proposed game controller with adaptive parameter has a better performance.

4.4.3 CASE 3: SCENARIO WITH HUMAN–MACHINE CONFLICT GOAL

The attention of human driver can be easily focused again after being reminded, but machine system will not discover its inappropriate decision under complex

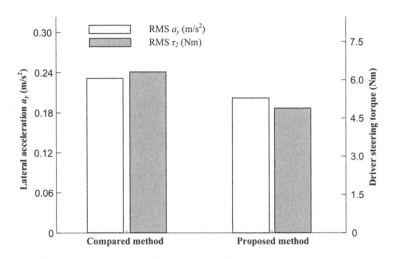

FIGURE 4.9   Performance analysis of different control approaches.

situation, such as driverless dilemma in [24], incorrect detection of sensory system. In this scenario, based on the different understanding of the complex traffic environment, the human driver and machine system are assumed to have conflict driving goals. The target path of human driver is a straight driving line along the road, and the target path of machine system is a lane change motion to avoid obstacle. The driver is arranged to maintain the original forward driving and resist the assistance from machine system. The experimental result is shown in Figure 4.10. With the proposed adaptive shared steering algorithm, a human-oriented game control can be realized by considering the real intention of human driver.

Based on the degree of conflict between two participants, an adaptive game parameter R_1 is adjusted according to the intentional action of human driver. The intention of driver is identified by the cumulative amplitude of steering torque. From the result in Figure 4.10, it can be found that the synthetic path is returned to straight driving line after a period of struggle. The synthetic path by the compared method is a compromise path which cannot achieve a flexible shared control in accordance with the driver intention. From the evaluation results of lateral acceleration and driver steering torque in Figure 4.11, the proposed game controller with adaptive parameter has a better control smoothness and comfort than the compared method. This kind of adaptive shared control framework have a human-oriented characteristic during the solution of decision conflict.

4.5   CONCLUSION

In this chapter, a game-based shared steering control for human–machine interaction is investigated in emergency scenarios. The shared control between the

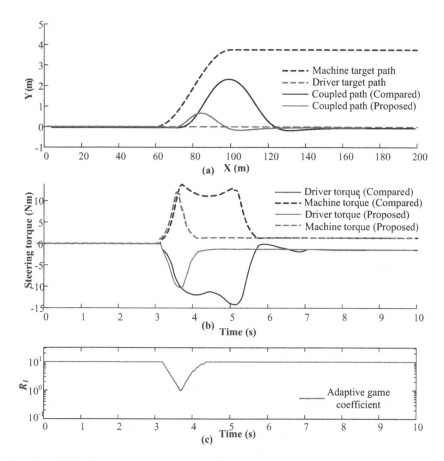

FIGURE 4.10 Experimental result when driver deliberately resists the assistance from machine system.

human driver and machine system is conducted by the coupling of steering torque on the handwheel. This kind of torque-based interaction mechanism can achieve a flexible game result from two participants. An adaptive parameter adjustment scheme is designed to achieve dynamic Nash equilibrium during game control. The shared steering is developed to obtain a reasonable game equilibrium by dynamically changing authority distribution coefficient. The proposed approach with the consideration of human intention can avoid single machine-oriented control results in which human plays a dominant role in emergency driving conditions.

The performance of game-based shared steering control is verified by the HIL experiments. From the results of different traffic scenarios, the proposed algorithm can effectively improve safety, and reduce the human–machine competition conflicts during the game control. The shared steering control in this study is focused on the

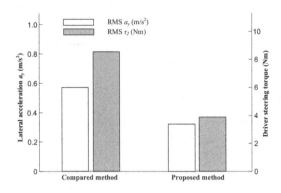

FIGURE 4.11   Performance analysis of different control approaches.

lateral dynamics, a game principle with consideration of longitudinal control will be studied in future.

REFERENCES

[1] Galvani, M. (2019). History and future of driver assistance. *IEEE Instrum. Meas. Mag.*, 22(1):11–6.

[2] Li, L., Wen, D., Zheng, N. N. and Shen, L. C. (2012). Cognitive cars: A new frontier for ADAS research. *IEEE Trans. Intell. Transp. Syst.*, 13(1):395–407.

[3] Hu, Y., Ting, Q. U., Liu, J. et al. (2019). Human-machine cooperative control of intelligent vehicle: Recent developments and future perspectives. *Acta Autom. Sin.*, 45(7): 1261–1280.

[4] Huang, M., Gao, W., Wang, Y., et al. (2019). Data-driven shared steering control of semi-autonomous vehicles. *IEEE Trans. Hum. Mach. Syst.*, 99:1–12.

[5] Han, J., Zhao, J., Zhu, B. and Song, D. (2022). Adaptive steering torque coupling framework considering conflict resolution for human-machine shared driving. *IEEE Trans. Intell. Transp. Syst.* 23(8): 10983–10995. doi: 10.1109/TITS.2021.3098466

[6] Wu, Y., Wei, H., Chen, X. et al. (2020). Adaptive authority allocation of human–automation shared control for autonomous vehicle. *Int. J. Autom. Technol.*, 21(3):541–553.

[7] Sentouh, C., Nguyen, A. T., Benloucif, M. A., et al. (2019). Driver-automation cooperation oriented approach for shared control of lane keeping assist systems. *IEEE Trans. Control Syst. Technol.*, 27(5):1962–1978.

[8] Guo, C., Sentouh, C., Popieul, J. C., et al. (2019). Predictive shared steering control for driver override in automated driving: A simulator study. *Transp. Res. Part F: Traffic Psychol. Behav.*, 61:326–336.

[9] Brandt, T., Sattel, T. and Bohm, M. (2007). Combining haptic human-machine inter-action with predictive path planning for lane-keeping and collision avoidance systems. *Intell. Veh. Symp.*, pages 582–587 IEEE.

[10] Mars, F., Deroo, M. and Hoc, J. M. (2014). Analysis of human-machine cooperation when driving with different degrees of haptic shared control. *IEEE Trans. Haptics*, 7(3):324–333.

[11] Canale, M., Fagiano, L. and Signorile, M. C. (2011). A model predictive control approach to vehicle yaw control using identified models. *Proc. Inst. Mech. Eng. D, J. Automob. Eng.*, 225:1475–1486.

[12] Falcone, P., Tseng, H. E., Borrelli, F., Asgari, J. and Hrovat, D. (2008). MPC–based yaw and lateral stabilization via active front steering and braking. *Veh. Syst. Dyn.*, Suppl. 46:611–628.

[13] Cole, D. J., Pick, A.J. and Odhams, A.M.C. (2006). Predictive and linear quadratic methods for potential application to modelling driver steering control. *Veh. Syst. Dyn.*, 44:259–284.

[14] Ji, X., Yang, K., Na, X., et al. (2019). Feedback game-based shared control scheme design for emergency collision avoidance: A fuzzy-LQR approach. *J. Dyn. Syst. Measure. Control*, 141(8):081005.

[15] Ji, X., Yang, K., Na, X. et al. (2019). Shared steering torque control for lane change assistance: A stochastic game-theoretic approach. *IEEE Trans. Ind. Electron.*, 66(4):3093–105.

[16] Na, X. and Cole, D. J. (2013). Linear quadratic game and non-cooperative predictive methods for potential application to modelling driverCAFS interactive steering control. *Veh. Syst. Dyn.*, 51(2):165–198.

[17] Na, X. and Cole, D. J. (2019). Modelling of a human driver s interaction with vehicle automated steering using cooperative game theory. *IEEE/CAA J. Autom. Sin.*, 6(5):1095–107.

[18] Flad, M., Otten, J., S. Schwab, S. et al. (2014). Steering driver assistance system: A systematic cooperative shared control design approach. *IEEE Int. Conf. Sys..* IEEE.

[19] Ma, B., Liu, Y., Na, X., Liu, Y., Yang, Y. (2019). A shared steering controller design based on steer-by-wire system considering human-machine goal consistency. *J. Frank. Inst.*, 356(8):4397–4419.

[20] Li, M., Song, X., Cao, H., et al. (2019). Shared control with a novel dynamic authority allocation strategy based on game theory and driving safety field. *Mech. Syst. Signal Process.*, 124:199–216.

[21] Tamaddoni, S. H., Ahmadian, M. and Taheri, S. (2011). Optimal vehicle stability control design based on preview game theory concept. *American Control Conference.* IEEE.

[22] Ludwig, J., Gote, C., Flad, M., et al. (2017). Cooperative dynamic vehicle control allocation using time-variant differential games. *2017 IEEE International Conference on Systems, Man and Cybernetics.* IEEE.

[23] Wu, X., Zhang, M., Xu, M. (2019). Active tracking control for steer-by-wire system with disturbance observer. *IEEE Trans. Veh. Technol.*, 68(6):5483–5493.

[24] Greene, J. D. (2016). Ethics. Our driverless dilemma. *Science*, 352(6293):1514–1515.

[25] Balachandran, A. and Gerdes, J. C. (2014). Designing steering feel for steer-by-wire vehicles using objective measures. *IEEE/ASME Trans. Mechatron.*, 20(1):373–383.

[26] Wu, X., Qiao, B. and Su, C (2020). Trajectory planning with time-variant safety margin for autonomous vehicle lane change. *Appl. Sci.*, 10(5):1626.

[27] Tan, Y., Shen, H., Huang, M., et al. (2016). Driver directional control using two–point preview and fuzzy decision. *J. Appl. Math. Mech.*, 80(6):459–465.

[28] Houenou, A., Bonnifait, P., Cherfaoui, V., et al. (2013). Vehicle trajectory prediction based on motion model and maneuver recognition. *IEEE/RSJ International Conference on Intelligent Robots and Systems.* IEEE.

[29] Wu, X.; Su, C.; Yan, L (2023). Human–Machine Shared Steering Control for Vehicle Lane Changing Using Adaptive Game Strategy. Machines, 11: 838.

5 Active Control of Four-Wheel Steering Vehicle

5.1 INTRODUCTION

5.1.1 BACKGROUND OF 4WS TECHNOLOGY

In recent years, with the improvement of people's living standards and the gradual growth of the city, transportation has received more and more attention from consumers, and thus the car plays an increasingly important role in people's daily lives. When it comes to choosing cars, people tend to focus on driving safety, maneuvering stability, and other aspects of performance indicators. As an important part of the vehicle, the function of the steering system is to adjust the driving direction, ensuring the vehicle's movement in line with the driver's intention for lateral control. Because of that, the steering system plays a very important role in enhancing the driver's comfort. The performance of the steering system is an important criterion for the vehicle's safety and reliability, and the development of the steering technology also marks the development of overall vehicle technology [1].

The steering system of traditional vehicles mostly adopts front-wheel steering (FWS). The driver gives the steering wheel a certain angle, then the mechanical device transmits to the front wheels for the corresponding rotation. Meanwhile, the rear wheels are subjected to the force of the ground and deform elastically, resulting in a follow-up steering. To ensure that the tires are in a pure rolling state, the turning center of the vehicle is located at the intersection of the front axle and the extension line of the rear axle. Because the rear wheel follow-up steering angle is very small, the center of the circle will only be changed by the front wheel steering angle, resulting in the vehicle's turning radius not adapting to the people's variable needs. This is directly reflected in the following working conditions: at low speeds, the vehicle needs a large steering wheel angle to achieve a small turning radius, affecting the driver's comfort; at high speeds, a small rotation of the steering wheel will lead to a large yaw rate, affecting the safety of driving the vehicle.

DOI: 10.1201/9781003481669-5

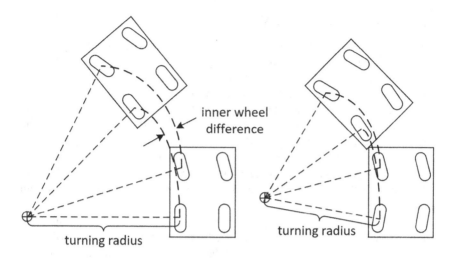

FIGURE 5.1 Comparison of front-wheel steering and four-wheel steering trajectories. Source: Figure created by authors.

As people's requirements for vehicle stability and safety improve, traditional vehicle steering technology has gradually failed to adapt to people's needs for vehicle maneuvering stability. Thanks to the development of electronic control technology, four-wheel steering technology is rapidly emerging. New active steering technologies, including active rear-wheel steering and active four-wheel steering, are rapidly evolving. As shown in Figure 5.1, at low speeds the rear wheels of the vehicle can rotate in the direction opposite to the front wheels, and the intersection of the lateral extension lines of the front and rear wheels is closer to the vehicle, thus reducing the turning radius and increasing the agility of low-speed turns. While at high speeds, the front and rear wheels of the vehicle rotate in the same direction, so that the intersection of the lateral extensions of the front and rear wheels is farther away from the vehicle, which increases the turning radius and increases the stability of high-speed turning.

With the rise and development of vehicle X-by-Wire technology, the wire-controlled chassis eliminates the mechanical connection device between the front wheels and the steering wheel and instead adopts the electrical signal to transmit steering instructions – so that the front and rear wheels can be steered independently of the steering wheel, which provides a convenient way of controlling the front and rear wheel steering angles. The active four-wheel steering technology relies on the steer-by-wire technology, directly controlling the steering angles of the front and rear wheels through the control algorithm so that the vehicle can adapt to the steering requirements of a variety of working conditions, and improve the stability of the vehicle's maneuvering. With the active control of steering a and front and rear wheel synergistic steering, active steering technologies such as active rear-wheel steering and active four-wheel steering have a broad application prospect.

5.1.2 Research Status of 4WS Technology

Four-wheel steering technology can be divided into rear-wheel follow-up steering technology and rear-wheel active steering technology. The early four-wheel steering technology was mostly rear-wheel follow-up steering technology. The rear wheels of this type of vehicle are connected to the steering wheel through mechanical devices, and the steering wheel directly controls the angle of the rear wheels through mechanical transmission. In 1907, Daimler AG of Germany developed a mechanically connected rear-wheel follow-up 4WS vehicle, which achieved good handling stability in the automotive market at that time. In 1991, Mitsubishi Corporation launched the 300GT sports car, which was equipped with a hydraulic 4WS system, improving the vehicle's steering performance at high speeds.

The rear-wheel active steering technology is different, with the rear-wheel follow-up steering, and it is not constrained by mechanical connections. Based on the by-wire control technology, the rear-wheel active steering is realized through the electronically controlled motor. Currently, the active four-wheel steering system has been widely used in 4WS vehicles. One of most representative products is the Quadrasteer system from the U.S. Delphi Corporation in 2000, which is shown in Figure 5.2. The Quadrasteer drives the rear wheels to make turning motions through a motor actuator on the rear axle of the vehicle. The Quadrasteer system can improve the vehicle's steering flexibility at low speeds and steering stability at high speeds.

Quadrasteer

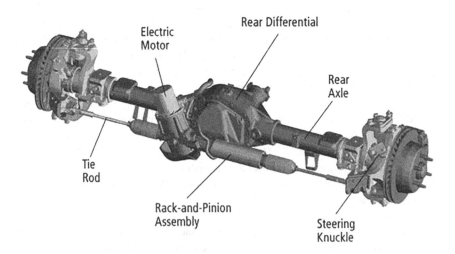

FIGURE 5.2   Four-wheel steering system of Quadrasteer.

Source: Hunting, Benjamin. *Driving Line*, Jan. 5, 2021.

5.2 ACTIVE CONTROL STRATEGIES OF MULTI-WHEEL STEERING VEHICLE

With the development of electronic control technology and automobile technology, people have higher and higher requirements for the driving stability of the vehicle's operation. One of these advanced control technologies is active steering technology. The main function of active steering technology is to make the car more flexible and stable during the turning process. This section mainly focuses on active steering technology, including active rear-wheel steering, fully active four-wheel steering strategy research with integrated front-wheel steering by wire. Moreover, through the designed vehicle dynamic model, vehicle side-slip angle estimation algorithms based on generalized Kalman filter observer are formulated for the active steering control system.

5.2.1 FRONT-WHEEL STEERING VEHICLE MODEL

In order to investigate the control algorithms of 4WS vehicle, a 2DOF vehicle dynamic model was established by means of linear simplification and parameter conformity. The diagrammatic sketch of the vehicle model is shown in Figure 5.3. Some factors that have little effect on the vehicle's lateral dynamics can be ignored. The basic assumptions for the simplified vehicle model are shown in the following:

(1) The influence of the suspension on the vehicle is not considered, that is, only the lateral motion along the y-axis and the swing around the z-axis are considered, and the pitch, roll, and vertical motion are not considered;

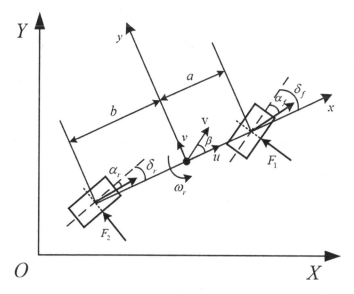

FIGURE 5.3 Linear 2DOF 4WS vehicle dynamics model.
Source: Figure created by authors.

(2) The vehicle speed is constant;
(3) The front and rear tire slip angles are small, and the cornering stiffness is always in the linear range;
(4) The effect of air resistance on the vehicle is ignored.

The resultant force F_y along the y-axis and the resultant moment M_z around the z-axis at the center of mass of the vehicle are shown below:

$$\begin{cases} F_y = F_1 \cos \delta_f + F_2 \cos \delta_r \\ M_z = aF_1 \cos \delta_f - bF_2 \cos \delta_r \end{cases} \tag{5.1}$$

Considering the small front and rear wheel angles δ_f and δ_r, it can be approximated as:

$$\cos \delta_f = \cos \delta_r = 1 \tag{5.2}$$

Therefore, Eqn. (5.1) can be written as:

$$\begin{cases} F_y = F_1 + F_2 \\ M_z = aF_1 - bF_2 \end{cases} \tag{5.3}$$

From the kinetic equation, Eqn. (5.3) can be further written as:

$$\begin{cases} F_y = F_1 + F_2 = ma_y \\ M_z = aF_1 - bF_2 = I_z \dot{\omega}_r \end{cases} \tag{5.4}$$

The expressions for the front and rear wheel cornering forces F_1 and F_2 are shown below:

$$\begin{cases} F_1 = k_1 \alpha_f \\ F_2 = k_2 \alpha_r \end{cases} \tag{5.5}$$

where k_1 and k_2 are the equivalent cornering stiffness of the front and rear axles.

Based on the motion of the vehicle, the expression for the slip angles can be obtained as shown below:

$$\begin{cases} \alpha_f = \dfrac{v + a\omega_r}{u} - \delta_f = \beta + \dfrac{a\omega_r}{u} - \delta_f \\ \alpha_r = \dfrac{v - b\omega_r}{u} - \delta_r = \beta - \dfrac{b\omega_r}{u} - \delta_r \end{cases} \tag{5.6}$$

The expression for the lateral acceleration a_y of the vehicle is shown below:

$$a_y = \frac{dv}{dt} + u\frac{d\theta}{dt} = \dot{v} + u\omega_r \tag{5.7}$$

Substituting Eqns. (5.5)(5.6)(5.7) into Eqn. (5.4) yields

$$\begin{cases} m(\dot{v} + u\omega_r) = (k_1 + k_2)\beta + \dfrac{1}{u}(ak_1 - bk_2)\omega_r - k_1\delta_f - k_2\delta_r \\ I_z\dot{\omega}_r = (ak_1 - bk_2)\beta + \dfrac{1}{u}(a^2k_1 + b^2k_2)\omega_r - ak_1\delta_f + bk_2\delta_r \end{cases} \tag{5.8}$$

Since the sideslip angle β is small, it can be assumed that $\tan\beta = \beta$, so

$$v = u\tan\beta = u\beta \tag{5.9}$$

Since the vehicle speed u is considered to be constant in the 2DOF model, it can be obtained by taking derivatives on both sides of the equation:

$$\dot{v} = u\dot{\beta} \tag{5.10}$$

Substituting Eqn. (5.10) into Eqn. (5.8) yields the linear 2DOF 4WS vehicle dynamics model as follows:

$$\begin{cases} mu(\dot{\beta} + \omega_r) = (k_1 + k_2)\beta + \dfrac{1}{u}(ak_1 - bk_2)\omega_r - k_1\delta_f - k_2\delta_r \\ I_z\dot{\omega}_r = (ak_1 - bk_2)\beta + \dfrac{1}{u}(a^2k_1 + b^2k_2)\omega_r - ak_1\delta_f + bk_2\delta_r \end{cases} \tag{5.11}$$

Translating the above equation into a state space model yields:

$$\begin{bmatrix} \dot{\beta} \\ \dot{\omega}_r \end{bmatrix} = \begin{bmatrix} \dfrac{k_1 + k_2}{mu} & \dfrac{ak_1 - bk_2}{mu^2} - 1 \\ \dfrac{ak_1 - bk_2}{I_z} & \dfrac{a^2k_1 + b^2k_2}{I_z u} \end{bmatrix} \begin{bmatrix} \beta \\ \omega_r \end{bmatrix} + \begin{bmatrix} -\dfrac{k_1}{mu} & -\dfrac{k_2}{mu} \\ -\dfrac{ak_1}{I_z} & \dfrac{bk_2}{I_z} \end{bmatrix} \begin{bmatrix} \delta_f \\ \delta_r \end{bmatrix} \tag{5.12}$$

5.2.2 Ideal Steering Reference Model

Previous research and practice have shown that drivers have the best driving experience with vehicles whose tires are in the linear zone and have appropriate understeer characteristics. Therefore, the relevant literature takes the steering characteristics of a conventional front-wheel steering vehicle in the linear region of the tire as the ideal steering characteristics pursued by the control strategy [2]. In this section, the control objectives of the active steering control strategy are designed to follow two points: keeping the sideslip angle of the vehicle around zero and tracking the ideal yaw velocity.

To keep the sideslip angle of the vehicle around zero, there are mainly several reasons:

(1) The velocity direction of the vehicle's center of mass should keep pace with the longitudinal axis direction, which is used to avoid tail sway and instability at the rear of vehicle during high-speed cornering motion.
(2) The drivers are provided with a better driving field of vision, and have better path planning result.
(3) The vehicle can achieve better traction performance and vehicle body attitude.

To track the yaw velocity for the ideal steering model of the vehicle, there are also several reasons:

(1) The steering sensitivity of the vehicle and the driver's handling performance can be similar to the conventional front-wheel steering vehicles, which greatly ensures the driving stability and safety.
(2) The yaw velocity can be controlled within a certain range, preventing the vehicle from having large under-steering or over-steering characteristics, and avoiding the dangerous results during the vehicle's steering control at high-speed obstacle avoidance.
(3) If the vehicle is operating in the linear range of the tires, the driver can achieve a better driving experience and vehicle handling stability.

Based on the above discussion, the control objective of the control algorithm is set to be consistent with the state parameters of a conventional front-wheel steering vehicle that is in the tire linear region, so that the ideal yaw velocity of the vehicle is

$$r_d = k_{rd}\delta_f^* = \frac{u/L}{1+Ku^2}\delta_f^* \qquad (5.13)$$

where K is the stability factor, and its expression is as follows:

$$K = \frac{m}{L^2}\left(\frac{a}{k_2} - \frac{b}{k_1}\right) \qquad (5.14)$$

For the ideal yaw velocity, it must also be determined whether the current road conditions will allow the vehicle to achieve it, which requires consideration of the limitations of the road adhesion coefficient. The maximum adhesion given to the tires by the road surface is as follows:

$$F_\mu = \mu mg \qquad (5.15)$$

This maximum adhesion force limits the lateral acceleration of the vehicle. If all this adhesion force is provided to the lateral motion of the vehicle, at this time the lateral acceleration of the vehicle is maximum. So, the lateral acceleration of the vehicle should satisfy the following constraints:

$$\left| a_y \right| \le \frac{F_\mu}{m} = \mu g \qquad (5.16)$$

The expression for the lateral acceleration of the vehicle at steady state is given below:

$$a_y = \frac{V^2}{R} = V \cdot r = u \cdot r \qquad (5.17)$$

Substituting Eqn. (5.17) into Eqn. (5.16) yields:

$$\left| r_d \right| \le \frac{\mu g}{u} \qquad (5.18)$$

Combining Eqn. (5.18), Eqn. (5.13) can be modified as:

$$r_d = \min\left\{ \left| \frac{u/L}{1 + Ku^2} \delta_f^* \right|, \left| \frac{\mu g}{u} \right| \right\} \mathrm{sgn}(\delta_f^*) \qquad (5.19)$$

If the ideal yaw velocity is only set to a constant value, it will lead to poor transient response. Phenomena such as overshooting and oscillating will appear, which affects the steering performance of the vehicle. Therefore, to improve the transient response process, a first-order lag element can be added. At the same time, the road surface adhesion coefficient limits the range of the ideal yaw velocity. Integrating the first-order lag element and road surface adhesion coefficient, we can obtain the ideal model of the yaw velocity as shown below:

$$r_d = \min\left\{ \left| \frac{u/L}{1 + Ku^2} \delta_f^* \right|, \left| \frac{\mu g}{u} \right| \right\} \mathrm{sgn}(\delta_f^*) \cdot \frac{1}{1 + \tau_r s} \qquad (5.20)$$

where τ_r is the lag element time constant and its empirical range is generally within 0.10~0.25. However, in extreme conditions such as high speed or low road adhesion coefficient, the lag element time constant can be appropriately adjusted smaller to maintain the overall stability of the vehicle.

Define the ideal sideslip angle as follows:

$$\beta_d = \frac{k_{\beta d}}{1 + \tau_\beta s} \delta_f^*$$ (5.21)

where $k_{\beta d}$ is the steady-state sideslip angle gain, τ_β is the lag element time constant. Since the ideal sideslip angle is set to zero, we can assume that $k_{\beta d} = 0$.

5.2.3 FRONT- AND REAR-WHEEL PROPORTIONAL STEERING CONTROL STRATEGY

In order to realize the vehicle flexibility at low speed and the stability at high speed, and at the same time consider the running time of the control algorithm and the design complexity, this section first builds an upper layer controller based on feed-forward steering control strategy. The lower layer performs four-wheel torque distribution on the obtained longitudinal force and additional yaw moment through two driving force distribution methods: average distribution and sequential quadratic programming.

The feed-forward front and rear wheel proportional steering control strategy is a traditional 4WS control strategy, and it is introduced to compare with the proposed active 4WS control strategy in the later section. Based on the proportional steering control strategy, the tire steering angle of the vehicle satisfies the following relationship:

$$\delta_r = K \cdot \delta_f$$ (5.22)

where K is the front and rear wheel ratio coefficient.

The control objective of this control strategy is to compress the sideslip angle to zero and to keep the yaw velocity unchanged when the vehicle is steering into a steady state [3]. Therefore, the following conditions can be obtained:

$$\begin{cases} \dot{\omega}_r = 0 \\ \beta = 0 \end{cases}$$ (5.23)

Combining the linear 2DOF 4WS vehicle dynamics model, substituting the above condition (5.23) and Eqn. (5.22) into Eqn. (5.11) yields:

$$\begin{cases} \dfrac{1}{u}(ak_1 - bk_2)\omega_r - mu\omega_r = k_1\delta_f + k_2\delta_r \\ \dfrac{1}{u}(a^2k_1 + b^2k_2)\omega_r = ak_1\delta_f - bk_2\delta_r \end{cases}$$ (5.24)

Eliminating ω_r from the above equation yields:

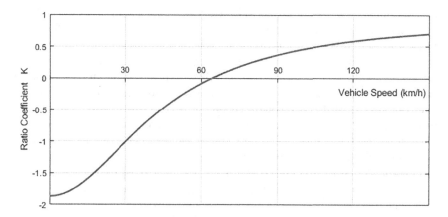

FIGURE 5.4 Relationship between the ratio coefficient K and the vehicle speed u. Source: Figure created by authors.

$$K = \frac{\delta_r}{\delta_f} = \frac{-b - \dfrac{mau^2}{k_2 L}}{a - \dfrac{mbu^2}{k_1 L}} \qquad (5.25)$$

Figure 5.4 shows the relationship between the ratio coefficient K and the vehicle speed u for the vehicle parameters (Specific vehicle parameters are introduced in Section 5.3).

5.2.4 ACTIVE REAR-WHEEL STEERING BY LQR CONTROLLER

LQR control is also known as linear quadratic optimal control. Optimal control refers to seeking an optimal control scheme under given constraints, combining the control objective and the controlled system, to achieve the desired control effect for performance indicators. LQR control can often be used for linear problems with multiple control objectives to optimize the overall performance of each control objective [4].

The control objective of LQR is to minimize the error between the state variable and the desired value, this error can be measured by the certain performance index, so LQR control is committed to minimizing the performance index. In LQR control, the performance index is defined as follows:

$$J = \frac{1}{2} X^T(t_f) S X(t_f) + \frac{1}{2} \int_{t_0}^{t_f} \left[X^T(t) Q X(t) + U^T(t) R U(t) \right] dt \qquad (5.26)$$

In the formula, S, Q, and R are the weighted positive definite matrices representing the weights of each item in the performance index. The first term of the performance index indicates the difference between the system state variables and the expected value at the terminal moment t_f. The second term indicates the cumulative tracking error between the system state variables and the expected value in the evaluation period from t_0 to t_f, which is used to evaluate the control effect; and the third term indicates the size of the total control energy in the evaluation period from t_0 to t_f, which is related to the size of the input variable $U(t)$. If the input variable is too large, it will consume a large amount of control energy and reduce the control performance.

Different from the feed-forward proportional steering control strategy, an upper layer controller based on LQR control is employed. Solving the Riccati equation in real time through an iterative method to obtain the rear wheel rotation angle and additional yaw moment, so as to complete the optimization of the side slip angle and yaw rate. The active rear-wheel steering LQR controller takes the reference front wheel angle input from the driver as the control input, and takes the vehicle rear wheel angle as the control output. It belongs to the single-input-single-output (SISO) system and can only take one as the control objective between tracking the ideal yaw velocity and compressing sideslip angle [5].

According to the vehicle dynamics model established in section 5.2.1, it can be obtained from Eqn. (5.12):

$$\dot{X} = AX + BU + CW \tag{5.27}$$

where

$$X = \begin{bmatrix} \beta \\ r \end{bmatrix}, \quad U = \begin{bmatrix} \delta_r \end{bmatrix}, \quad W = \begin{bmatrix} \delta_f \end{bmatrix}$$

$$A = \begin{bmatrix} \dfrac{k_1 + k_2}{mu} & \dfrac{ak_1 - bk_2}{mu^2} - 1 \\ \dfrac{ak_1 - bk_2}{I_z} & \dfrac{a^2 k_1 + b^2 k_2}{I_z u} \end{bmatrix}, \quad B = \begin{bmatrix} -\dfrac{k_2}{mu} \\ \dfrac{bk_2}{I_z} \end{bmatrix}, \quad C = \begin{bmatrix} -\dfrac{k_1}{mu} \\ -\dfrac{ak_1}{I_z} \end{bmatrix}$$

The ideal reference model for a vehicle is defined as:

$$X_d = A_d \cdot \delta_f \tag{5.28}$$

where

$$X_d = \begin{bmatrix} \beta_d \\ r_d \end{bmatrix}, \quad A_d = \begin{bmatrix} 0 \\ \dfrac{u/L}{1 + Ku^2} \end{bmatrix}$$

In the equation, β_d is the ideal sideslip angle, and r_d is the steady-state yaw rate of the front wheel steering vehicle.

For the active rear-wheel steering LQR control, The first term of the performance index J has little effect on the control and can be ignored, so the performance index is shown as follows:

$$J = \frac{1}{2}\int_{t_0}^{t_f}\left[(X - X_d)^T Q(X - X_d) + U^T R U\right]dt \qquad (5.29)$$

where

$$Q = \begin{bmatrix} q_\beta & 0 \\ 0 & q_r \end{bmatrix}, \ R = r_{\delta_r}$$

In the equation, q_β, q_r, and r_{δ_r} respectively represent the importance that the control algorithm places on the sideslip angle, yaw rate, and rear wheel angle.

Construct the Hamiltonian function as follows:

$$H = \frac{1}{2}(X - X_d)^T Q(X - X_d) + \frac{1}{2}U^T R U + \lambda^T(t)(AX + BU + CW) \quad (5.30)$$

The control equation is:

$$\frac{\partial H}{\partial U} = RU + B^T \lambda(t) = 0 \qquad (5.31)$$

that is,

$$U = -R^{-1}B^T \lambda(t) \qquad (5.32)$$

The adjoint equation is:

$$\dot{\lambda}(t) = -\frac{\partial H}{\partial e} = -\frac{\partial H}{\partial(X - X_d)} = -Q(X - X_d) - A^T\lambda(t) \qquad (5.33)$$

We can assume that:

$$\lambda(t) = P(t)X - \xi(t) \qquad (5.34)$$

Derived:

$$\dot{\lambda}(t) = \dot{P}(t)X + P(t)\dot{X} - \dot{\xi}(t) \qquad (5.35)$$

Since the system is a linear constant system, A, B, Q, and R are constant matrices. So when $t \to \infty$, there is:

$$\dot{P}(t) = 0, \ \dot{\xi}(t) = 0 \tag{5.36}$$

Hence,

$$\dot{\lambda}(t) = P(t)\dot{X} \tag{5.37}$$

Substituting Eqns. (5.27)(5.32)(5.34)(5.37) into Eqn. (5.33) and simplifying yields:

$$(Q + PA + A^T P - PBR^{-1}B^T P)X = (A^T - PBR^{-1}B^T)\xi(t) + (QA_d - PC)W \tag{5.38}$$

From the above equation, the algebraic Riccati equation is obtained as follows:

$$Q + PA + A^T P - PBR^{-1}B^T P = 0 \tag{5.39}$$

$$(A^T - PBR^{-1}B^T)\xi(t) + (QA_d - PC)W = 0 \tag{5.40}$$

The value of $P(t)$ can be solved from Eqn. (5.39).

From Eqn. (5.40), the expression for $\xi(t)$ can be solved as follows:

$$\xi(t) = (A^T - PBR^{-1}B^T)^{-1}(PC - QA_d)W \tag{5.41}$$

Substituting Eqns. (5.34)(5.41) into Eqn. (5.32) yields:

$$U = \delta_r = -R^{-1}B^T PX + R^{-1}B^T(A^T - PBR^{-1}B^T)^{-1}(PC - QA_d)W \tag{5.42}$$

That is, the control equation is obtained as shown below:

$$U = \delta_r = U_{eq} + U_{nl} \tag{5.43}$$

where

$$\begin{cases} U_{eq} = k_1 X \\ U_{nl} = k_2 \delta_f \end{cases}$$

$$k_1 = -R^{-1}B^T P, \ k_2 = R^{-1}B^T(A^T - PBR^{-1}B^T)^{-1}(PC - QA_d)$$

The active rear-wheel steering LQR control model is mainly divided into three main parts: the weighting coefficient calculation module, the tire linear zone

LQR controller, and the tire nonlinear zone LQR controller. Among them, the parameter q_β in the Q matrix of the tire linear zone LQR controller is small, while the parameter q_β in the Q matrix of the tire nonlinear zone LQR controller is large. A weighting coefficient λ related to the tire slip angle is introduced to weigh these two controllers to obtain the final decision-making rear wheel steering angle. Thus, to a certain extent, the bad influence of tire cornering stiffness nonlinearity on the performance of the control model is weakened.

5.2.5 ACTIVE FOUR-WHEEL STEERING BY LQR CONTROLLER

Unlike active rear-wheel steering, the active four-wheel steering LQR controller calculates the ideal front and rear wheel angles of the vehicle from the reference front wheel angles output from the driver model. It belongs to the multi-input-multi-output (MIMO) system, and it can accomplish the two control objectives of tracking the ideal yaw rate and compressing the sideslip angle at the same time [6].

According to the vehicle dynamics model established in section 5.2.1, it can be obtained from Eqn. (5.12):

$$\dot{X} = AX + BU \tag{5.44}$$

where

$$X = \begin{bmatrix} \beta \\ r \end{bmatrix}, \quad U = \begin{bmatrix} \delta_f \\ \delta_r \end{bmatrix}$$

$$A = \begin{bmatrix} \dfrac{k_1 + k_2}{mu} & \dfrac{ak_1 - bk_2}{mu^2} - 1 \\ \dfrac{ak_1 - bk_2}{I_z} & \dfrac{a^2 k_1 + b^2 k_2}{I_z u} \end{bmatrix}, \quad B = \begin{bmatrix} -\dfrac{k_1}{mu} & -\dfrac{k_2}{mu} \\ -\dfrac{ak_1}{I_z} & \dfrac{bk_2}{I_z} \end{bmatrix}$$

The ideal reference model for a vehicle is defined as:

$$\dot{X}_d = A_d X_d + B_d U_d \tag{5.45}$$

where

$$X_d = \begin{bmatrix} \beta_d \\ r_d \end{bmatrix}, \quad U_d = [\delta_f^*], \quad A_d = \begin{bmatrix} -\dfrac{1}{\tau_\beta} & 0 \\ 0 & -\dfrac{1}{\tau_r} \end{bmatrix}, \quad B_d = \begin{bmatrix} \dfrac{k_{\beta d}}{\tau_\beta} \\ \dfrac{k_{rd}}{\tau_r} \end{bmatrix}$$

$$k_{\beta d} = 0, \quad k_{rd} = \frac{u/L}{1 + Ku^2}, \quad K = \frac{m}{L^2}\left(\frac{a}{k_2} - \frac{b}{k_1}\right)$$

In the equation, δ_f^* is the reference front wheel angle, τ_β and τ_r are taken as 0.1 for normal operating conditions and can be reduced to 0.05 for extreme operating conditions.

Define the tracking error as follows:

$$e = X - X_d = \begin{bmatrix} \beta - \beta_d \\ r - r_d \end{bmatrix} \tag{5.46}$$

Coupling the derivation of the tracking error e and Eqns. (5.44) (5.45) yields:

$$\dot{e} = \dot{X} - \dot{X}_d = AX + BU - A_d X_d - B_d U_d = \\ Ae + (A - A_d)X_d + BU - B_d U_d \tag{5.47}$$

Performance index J is defined as follows:

$$J = \frac{1}{2}\int_{t_0}^{t_f} \left[e^T Q e + U^T R U\right] dt \tag{5.48}$$

where

$$Q = \begin{bmatrix} q_\beta & 0 \\ 0 & q_r \end{bmatrix}, \quad R = \begin{bmatrix} r_{\delta_f} & 0 \\ 0 & r_{\delta_r} \end{bmatrix}$$

LQR control seeks the optimal control input $U(t)$ to achieve the minimum performance index J. Construct the Hamiltonian function as follows:

$$H = \frac{1}{2}e^T Q e + \frac{1}{2}U^T R U + \lambda^T(t)\dot{e} \tag{5.49}$$

Substituting Eqn. (5.47) into the above equation yields:

$$H = \frac{1}{2}e^T Q e + \frac{1}{2}U^T R U + \lambda^T(t)\left[Ae + (A - A_d)X_d + BU - B_d U_d\right] \tag{5.50}$$

The control equation is:

$$\frac{\partial H}{\partial U} = RU + B^T \lambda(t) = 0 \tag{5.51}$$

that is,

$$U = -R^{-1}B^T \lambda(t) \tag{5.52}$$

The adjoint equation is:

$$\dot{\lambda}(t) = -\frac{\partial H}{\partial e} = -Qe - A^T \lambda(t) \tag{5.53}$$

We can assume that:

$$\lambda(t) = P(t)e - \xi(t) \tag{5.54}$$

Derived:

$$\dot{\lambda}(t) = \dot{P}(t)e + P(t)\dot{e} - \dot{\xi}(t) \tag{5.55}$$

Since the system is a linear constant system, A, B, Q, and R are constant matrices. So when $t \to \infty$, there is:

$$\dot{P}(t) = 0, \; \dot{\xi}(t) = 0 \tag{5.56}$$

Hence,

$$\dot{\lambda}(t) = P(t)\dot{e} \tag{5.57}$$

Substituting Eqns. (5.47)(5.52)(5.54)(5.57) into Eqn. (5.53) and simplifying yields:

$$(Q + PA + A^T P - PBR^{-1}B^T P)e = (A^T - PBR^{-1}B^T)\xi(t) - P(A - A_d)X_d + PB_d U_d \tag{5.58}$$

From the above equation, the algebraic Riccati equation is obtained as follows:

$$Q + PA + A^T P - PBR^{-1}B^T P = 0 \tag{5.59}$$

$$(A^T - PBR^{-1}B^T)\xi(t) - P(A - A_d)X_d + PB_d U_d = 0 \tag{5.60}$$

The value of $P(t)$ can be solved from Eqn. (5.59).

From Eqn. (5.60), the expression for $\xi(t)$ can be solved as follows:

$$\xi(t) = (A^T - PBR^{-1}B^T)^{-1}\left[P(A - A_d)X_d - PB_d U_d \right] \tag{5.61}$$

Substituting Eqns. (5.54)(5.61) into Eqn. (5.52) yields:

$$U = \begin{bmatrix} \delta_f \\ \delta_r \end{bmatrix} = -R^{-1}B^T Pe + R^{-1}B^T(A^T - PBR^{-1}B^T)^{-1}P(A - A_d)X_d - \qquad (5.62)$$
$$R^{-1}B^T(A^T - PBR^{-1}B^T)^{-1}PB_dU_d$$

That is, the control equation is obtained as shown below:

$$U = \begin{bmatrix} \delta_f \\ \delta_r \end{bmatrix} = U_{eq1} + U_{eq2} + U_{nl} \qquad (5.63)$$

where

$$\begin{cases} U_{eq1} = k_1 e \\ U_{eq2} = k_2 X_d \\ U_{nl} = k_3 U_d \end{cases}$$

$$k_1 = -R^{-1}B^T P$$
$$k_2 = R^{-1}B^T(A^T - PBR^{-1}B^T)^{-1}P(A - A_d)$$
$$k_3 = -R^{-1}B^T(A^T - PBR^{-1}B^T)^{-1}PB_d$$

The structure of the active four-wheel steering LQR controller is shown in Figure 5.5, where the driver model outputs the reference front wheel angle δ_f^* based on closed-loop tracking of target trajectory or open-loop steering commands. The ideal steering reference model calculates the ideal yaw rate and ideal sideslip angle based on δ_f^* and vehicle speed signal u, and combines them to form the ideal vehicle state parameter X_d. The active four-wheel steering LQR controller takes

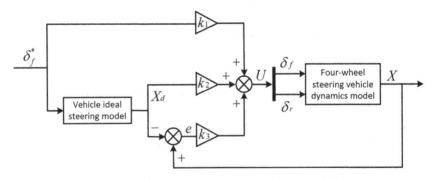

FIGURE 5.5   Active four-wheel steering LQR controller structure diagram.

Source: Figure created by authors.

the reference front wheel angle δ^{*}_{f}, vehicle state parameter X, and ideal vehicle state parameter X_{d} as inputs, and tracks the ideal yaw rate and ideal sideslip angle as control objectives. The front and rear wheel angles U are calculated and output to the vehicle dynamics model. Then the vehicle dynamics model calculates the vehicle state parameters and feeds them back to the active four-wheel steering LQR controller, completing the closed-loop of the system.

5.2.6 ACTIVE FOUR-WHEEL STEERING BY THREE-STEP CONTROLLER

Based on the discussion of LQR active four-wheel steering, a three-step steering control algorithm to achieve the ideal steering reference model is considered here. The reference of the ideal steering model is designed to keep the sideslip angle of the vehicle around zero and track the ideal yaw velocity.

The standard expression for a three-step controller is:

$$u = u_{s}(x) + u_{f}(x) + u_{e}(x) \qquad (5.64)$$

The design of the three-step controller for active four-wheel steering refers to the vehicle dynamics model in section 5.2.1. The detailed design steps of the three-step controller are introduced below:

(1) Steady-state-like control law $u_{s}(x)$

The purpose of the steady-state-like control law $u_{s}(x)$ is to obtain the output $y_{s}(x)$ in a steady state, but under this control law, the steady-state output is not effectively reached at transient time due to time constraints, so it is called steady-state-like control.

Referring to Eqn. (5.44), the design of the steady-state-like control law u_{s} satisfies the following equation:

$$\dot{X} = AX + Bu_{s} \qquad (5.65)$$

Since at steady state $\dot{X} = 0$, i.e., $\dot{\beta} = 0$, $\dot{r} = 0$. From Eqn. (5.65), it can be concluded that:

$$AX + Bu_{s} = 0 \qquad (5.66)$$

The solution to the steady-state-like control law is as follows:

$$u_{s} = -B^{-1}AX \qquad (5.67)$$

(2) Dynamic feed-forward control law $u_{f}(x)$

The feed-forward term $u_f(x)$ is used to correct the steady-state-like control law $u_s(x)$ to $u = u_s(x) + u_s(x)$. The dynamic change process of the control system is taken into account, which improves the dynamic performance of the system.

Referring to Eqn. (5.44), the dynamic feed-forward control law u_f is designed to satisfy the following equation:

$$\dot{X} = AX + B(u_s + u_f) \tag{5.68}$$

To track the ideal model, refer to Eqn. (5.45), we can assume that $\dot{X} = \dot{X}_d$, while substituting Eqn. (5.66) into Eqn. (5.68) can be obtained:

$$\dot{X}_d = Bu_f \tag{5.69}$$

The dynamic feed-forward control law is solved as follows:

$$u_f = B^{-1}\dot{X}_d \tag{5.70}$$

(3) State-dependent error feedback control law $u_e(x)$

Due to the influence of system uncertainty, the above two control laws still can not make the system achieve the desired control effect, so it is necessary to design the error feedback control law.

Define tracking error

$$e = y^*(x) - y(x) \tag{5.71}$$

Then the state-dependent error feedback control law can be designed in a PID-based structure:

$$u_e(x) = k_p e + k_i \int e dt + k_d \dot{e} \tag{5.72}$$

Referring to Eqn. (5.44), the state-dependent error feedback control law u_e is designed to satisfy the following equation:

$$\dot{X} = AX + B(u_s + u_f + u_e) \tag{5.73}$$

Define the tracking error as follows:

$$e = X_d - X \tag{5.74}$$

Substituting Eqns. (5.66)(5.69) into Eqn. (5.73) yields:

$$Bu_e = \dot{X} - \dot{X}_d = -\dot{e} \tag{5.75}$$

that is

$$\dot{e} = -Bu_e \qquad (5.76)$$

Design the error feedback control law u_e as a gain of the tracking error e as follows:

$$u_e = Me \qquad (5.77)$$

where

$$B = \begin{bmatrix} -\dfrac{k_1}{mu} & -\dfrac{k_2}{mu} \\ -\dfrac{ak_1}{I_z} & \dfrac{bk_2}{I_z} \end{bmatrix} = \begin{bmatrix} b_1 & b_2 \\ b_3 & b_4 \end{bmatrix}, \quad M = \begin{bmatrix} m_1 & m_2 \\ m_3 & m_4 \end{bmatrix}$$

Substituting Eqn. (5.77) into Eqn. (5.76) yields:

$$\dot{e} = -Bu_e = -BMe = -\begin{bmatrix} b_1 m_1 + b_2 m_3 & b_1 m_2 + b_2 m_4 \\ b_3 m_1 + b_4 m_3 & b_3 m_2 + b_4 m_4 \end{bmatrix} e \qquad (5.78)$$

To select the appropriate control parameter M, the above equation is decoupled. Assuming that M is a diagonal matrix that satisfies the following equation:

$$b_3 m_1 + b_4 m_3 = b_1 m_2 + b_2 m_4 = 0 \qquad (5.79)$$

Therefore

$$\begin{cases} \dfrac{m_1}{m_3} = -\dfrac{b_4}{b_3} = -\dfrac{bk_2/I_z}{-ak_1/I_z} = \dfrac{bk_2}{ak_1} \\ \dfrac{m_2}{m_4} = -\dfrac{b_2}{b_1} = -\dfrac{-k_2/mu}{-k_1/mu} = -\dfrac{k_2}{k_1} \end{cases} \qquad (5.80)$$

To satisfy the above equation, let us assume that the parameters k_a, k_b exist, making the following equation hold:

$$\begin{cases} m_1 = \dfrac{b}{(a+b)k_1} k_a \\ m_3 = \dfrac{a}{(a+b)k_2} k_a \end{cases} \quad \begin{cases} m_2 = \dfrac{1}{(a+b)k_1} k_b \\ m_4 = -\dfrac{1}{(a+b)k_2} k_b \end{cases} \qquad (5.81)$$

Referring to the form of matrix B, the above equation can be modified as:

$$\begin{cases} m_1 = \dfrac{mub}{(a+b)k_1}k_a & m_2 = \dfrac{I_z}{(a+b)k_1}k_b \\[4mm] m_3 = \dfrac{mua}{(a+b)k_2}k_a & m_4 = -\dfrac{I_z}{(a+b)k_2}k_b \end{cases} \tag{5.82}$$

Therefore, the expression for the state-dependent error feedback control law u_e is as follows:

$$u_e = Me = \begin{bmatrix} \dfrac{mub}{(a+b)k_1}k_a & \dfrac{I_z}{(a+b)k_1}k_b \\[4mm] \dfrac{mua}{(a+b)k_2}k_a & -\dfrac{I_z}{(a+b)k_2}k_b \end{bmatrix} e \tag{5.83}$$

where k_a, k_b are control parameters.

In summary, based on the above three control laws, the comprehensive control law of the active four-wheel steering three-step controller is:

$$u = u_s + u_f + u_e = -B^{-1}AX + B^{-1}\dot{X}_d + \begin{bmatrix} \dfrac{mub}{(a+b)k_1}k_a & \dfrac{I_z}{(a+b)k_1}k_b \\[4mm] \dfrac{mua}{(a+b)k_2}k_a & -\dfrac{I_z}{(a+b)k_2}k_b \end{bmatrix} e \tag{5.84}$$

The structure and model of the active four-wheel steering three-step controller are shown in Figure 5.6, respectively.

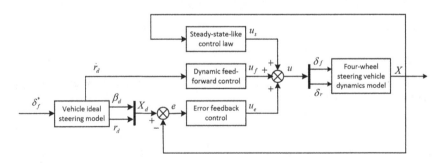

FIGURE 5.6   Active four-wheel steering three-step controller structure diagram.

Source: Figure created by authors.

The three-step controller performs steady-state-like control through the vehicle state parameter X, dynamic feed-forward control through the derivative of the vehicle ideal state parameter X_d, and error feedback control through the tracking error e between the vehicle ideal state parameter X_d and the vehicle current state parameter X. Combining these three sub-control modules, the front and rear wheel angles δ_f, δ_r are accumulated and output to the vehicle dynamics model. Then the vehicle dynamics model calculates the vehicle state parameters and feeds it back to the active four-wheel steering three-step controller, completing the closed-loop of the system.

5.2.7 ACTIVE FOUR-WHEEL STEERING SLIDING MODE CONTROLLER

As shown in Figure 5.7, the basic principle of sliding mode control (SMC) is to make the system reach the designed sliding mode surface through process 1 for a limited time, and then slide the system through process 2 to the stable state at point C (i.e. origin) through the control effect of the sliding mode surface, in which process 1 is called the converging mode, and process 2 becomes the sliding mode.

Figure 5.8 shows the process of system state changes in a real system under sliding mode control, which can result in system chattering. There are some special optimization methods for the chattering phenomenon, such as the fuzzy control method, approaching law, quasi sliding mode method, and so forth [7].

Define the tracking error between the system and the ideal model as follows:

$$e = X_d - X = \begin{bmatrix} \beta_d - \beta \\ r_d - r \end{bmatrix} \tag{5.85}$$

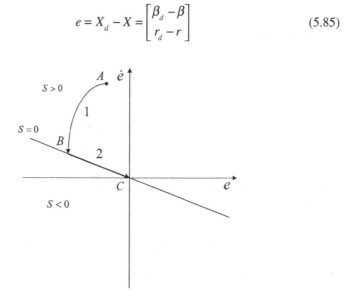

FIGURE 5.7 SMC principle schematic.
Source: Figure created by authors.

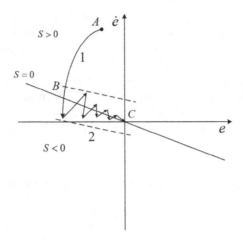

FIGURE 5.8   SMC schematic of the actual system.

Source: Figure created by authors.

Taking the derivative of the tracking error e and substituting Eqns. (5.44)(5.45) into it can obtain:

$$\dot{e} = \dot{X}_d - \dot{X} = A_d X_d + B_d U_d - AX - BU = \\ A_d e + (A_d - A)X + B_d U_d - BU$$

(5.86)

Design the sliding mode surface as:

$$S = Ce$$

(5.87)

where, C is the diagonal matrix:

$$C = \begin{bmatrix} c_1 & 0 \\ 0 & c_2 \end{bmatrix}$$

Taking the derivative of the switching function S and substituting Eqn. (5.86) into it yields:

$$\dot{S} = C\dot{e} = C\left[A_d e + (A_d - A)X + B_d U_d - BU \right]$$

(5.88)

The exponential approaching law is used as follows:

$$\dot{S} = -\varepsilon \cdot sign(S) - kS$$

(5.89)

The sliding mode control law for active four-wheel steering can be obtained by combining the Eqn. (5.88) and (5.89) as follows:

$$U = B^{-1}\left[A_d e + (A_d - A)X + B_d U_d\right] + B^{-1}C^{-1}\left[\varepsilon \cdot sign(S) + kS\right] \quad (5.90)$$

Due to the characteristics of the sign function, the chattering phenomenon occurs in the control process and the control is unstable. To solve the chattering problem of sliding mode control, the sign function in the approaching law is replaced by the saturation function:

$$sat(S) = \begin{cases} 1, & S > p \\ S/p, & |S| \le p \\ -1, & S < -p \end{cases} \quad (5.91)$$

where p denotes the thickness of the boundary layer and $p > 0$. If the thickness of the boundary layer is too thin, the suppression of chattering by the saturation function will be reduced. If the thickness of the boundary layer is too thick, the system state response will be too slow and the steady state error will become large. So the parameter p needs to be selected accurately.

Combined with the saturation function, the modified active four-wheel steering sliding mode control law is shown below:

$$U = B^{-1}\left[A_d e + (A_d - A)X + B_d U_d\right] + B^{-1}C^{-1}\left[\varepsilon \cdot sat(S) + kS\right] \quad (5.92)$$

The structure of the active four-wheel steering sliding mode controller is shown in Figure 5.9. The sliding mode controller receives the vehicle state parameter X, the reference front wheel angle δ^*_f, and the tracking error e between the ideal vehicle state parameter X_d and the current vehicle state parameter X, calculates the sliding mode surface, and uses the exponential approaching law containing saturation function for sliding mode control to obtain the front and rear wheel angles δ_f and δ_r, which are output to the vehicle dynamics model. Then the vehicle dynamics model calculates the vehicle state parameters and feeds them back to the

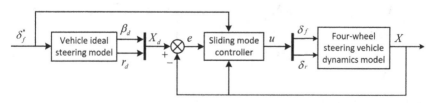

FIGURE 5.9 Active four-wheel steering sliding mode controller structure diagram.
Source: Figure created by authors.

active four-wheel steering sliding mode controller, completing the closed-loop of the system.

5.3 ANALYSIS OF ACTIVE CONTROL STRATEGIES OF MULTI-WHEEL STEERING VEHICLE

5.3.1 MODELLING THE SIMULATION AND TESTING ENVIRONMENT

Due to the complexity of the vehicle dynamics system, it is necessary to build a multi-DOF vehicle dynamics model. In this section, a simplified 8-DOF 4WS vehicle dynamics model is established for vehicle dynamics simulation and analysis, including longitudinal, lateral, yaw, roll, and four-wheel rotation degrees of freedom, as shown in the Figure 5.10.

(1) Longitudinal dynamics equation

$$m(\dot{u} - vr) - m_s h\dot{r}\varphi = \sum F_{xi} - \sum F_{fi} - F_w \tag{5.93}$$

where u and v are the longitudinal and lateral speeds of the vehicle, m_s is the spring-loaded mass, φ is the roll angle, h is the vertical distance from the center of the

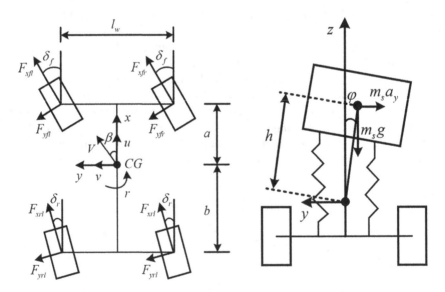

FIGURE 5.10 4WS vehicle dynamics model schematic.

Source: Figure created by authors.

spring-loaded mass to the roll axis, F_{xi} is the longitudinal tire force, F_{fi} is the rolling resistance of the tire, and F_w is the longitudinal air resistance.

(2) Lateral dynamics equation

$$m(\dot{v} + ur) + m_s h \ddot{\varphi} = \sum F_{yi} =$$
$$(F_{yfl} + F_{yfr})\cos \delta_f + (F_{yrl} + F_{yrr})\cos \delta_r + \qquad (5.94)$$
$$(F_{xfl} + F_{xfr})\sin \delta_f + (F_{xrl} + F_{xrr})\sin \delta_r$$

where δ_f, δ_r are the front and rear wheel steering angles, respectively.

(3) Yaw dynamics equation

$$I_z \dot{r} - I_{xz} \ddot{\varphi} = \sum M_{zi} =$$
$$a\left[(F_{xfl} + F_{xfr})\sin \delta_f + (F_{yfl} + F_{yfr})\cos \delta_f \right] -$$
$$b\left[(F_{xrl} + F_{xrr})\sin \delta_r + (F_{yrl} + F_{yrr})\cos \delta_r \right] + \qquad (5.95)$$
$$\frac{l_w}{2}\left[(F_{xfl} - F_{xfr})\cos \delta_f + (F_{yfr} - F_{yfl})\sin \delta_f \right] +$$
$$\frac{l_w}{2}\left[(F_{xrl} - F_{xrr})\cos \delta_r + (F_{yrr} - F_{yrl})\sin \delta_r \right]$$

where M_{zi} is the yaw moment on the vehicle and l_w is the wheelbase.

(4) Roll dynamics equation

$$I_x \ddot{\varphi} - I_{xz} \dot{r} + m_s h(\dot{v} + ur) = \sum M_{xi} = -C_\varphi \dot{\varphi} - K_\varphi \varphi + m_s gh\varphi \qquad (5.96)$$

where M_{xi} is the roll moment on the vehicle, $C\varphi$ and $K\varphi$ represent the roll damping and roll stiffness of the vehicle.

(5) Wheel rotation dynamics equation

$$I_w \dot{\omega}_i = T_{di} - T_{bi} - F_{xi} R \qquad (5.97)$$

where T_{di} and T_{bi} are the driving torque and braking torque of the i wheel, R is the rolling radius of the wheel, and I_w is the inertia of the wheel.

(6) Tire model

The dynamic characteristics of vehicles are mainly described through the vehicle dynamics equations mentioned earlier, in which most of the forces and moments received by the vehicle are transmitted to the body through the ground. As an indispensable part of a vehicle, tires are used to transmit the interaction between the ground and the vehicle body, allowing the vehicle to receive the driving force, braking force, lateral force, and so forth, provided by the ground, thereby changing the vehicle's motion. From this, it can be seen that tires have a significant impact on the motion of vehicles, which is related to the authenticity of the vehicle dynamics model. Due to the good estimation effect of the Magic Formula on tire force and its simple form, we apply the Magic Formula to establish the tire model for vehicles.

To fit the relationship curve between the lateral force exerted on the actual tire and the slip angle, the Magic Formula is shown below:

$$\begin{cases} F_y(\alpha_y) = D\sin\left\{C\arctan\left[B\alpha_y - E(B\alpha_y - \arctan(B\alpha_y))\right]\right\} + S_V \\ \alpha_y = \alpha + S_H \end{cases} \quad (5.98)$$

where, α is the tire slip angle and F_y is the tire lateral force, B, C, D, E are coefficients.

$$B = BCD / CD, C = b_0, D = b_1 F_z^2 + b_2 F_z, BCD =$$
$$(b_3 F_z^2 + b_4 F_z)*e^{-b_5 F_z}, E = b_8 F_z^2 + b_6 F_z + b_7$$

We use the '215/55 R17' tire model, and its parameter fitting results are shown in Table 5.1. According to the fitting curve, the equivalent cornering stiffness in the linear zone of the front and rear axle tires are −144070N/rad and −82396N/rad. The parameters of the established vehicle model are summarized in Table 5.2.

To verify the superiority of the LQR active steering control strategy over the conventional FWS and feed-forward proportional steering control strategies, as well as to explore the performance differences among these strategies, the steering performance of the vehicle under various operating conditions is tested by using the joint CarSim-Simulink simulation. The joint simulation mechanism is as follows: CarSim conveys the steering wheel angle and vehicle state parameters to Simulink. The steering control model in Simulink obtains the reference front wheel angle according to the steering wheel angle, which is then combined with the vehicle state parameters to calculate the four wheel angles of the vehicle. At last, the controller outputs it to CarSim for steering control to complete the closed loop.

TABLE 5.1
Fitting Values of Tire Model Parameters

Parameters	b_0	b_1	b_2	b_3	b_4	b_5	b_6	b_7	b_8
fitting values	1.414	-9.635	1021	-532.8	18150	1.74e-2	0.049	-0.465	2.8e-3

Source: Table created by authors.

TABLE 5.2
Vehicle Model Parameters

Parameters	Value	Parameters	Value
Mass m	$1413kg$	front track l_w	$1.675m$
front wheelbase a	$1.015m$	rear track l_w	$1.675m$
rear wheelbase b	$1.895m$	front cornering stiffness k_1	$-144070N\backslash rad$
Wheelbase L	$2.91m$	rear cornering stiffness k_2	$-82396N\backslash rad$
roll inertia I_x	$536.6kg.m^2$	steering ratio i	18.6
pitch inertia I_y	$1536.7kg.m^2$	wheel rolling radius R	$0.325m$
yaw inertia I_z	$1536.7kg.m^2$	wheel rotational inertia I_w	$1.5kg.m^2$
height of centroid h_g	0.54	vehicle windward area A	$2.2m^2$

Source: Table created by authors.

5.3.2 ANALYSIS OF LQR ACTIVE STEERING CONTROL METHODS

In this section, we will verify the effectiveness of the LQR control strategy in vehicle steering control compared to conventional FWS and feed-forward control strategy. The simulation experiments are conducted with the step steering control, lateral wind control, and double lane change control respectively.

5.3.2.1 Step Steering Simulation Test

The steering wheel angle step test is the basic test to evaluate the transient response of vehicle handling stability. During the test, the state parameters of the vehicle will be changed, mainly including the yaw velocity, sideslip angle, lateral acceleration, and so forth. The evaluation indexes are the transient response characteristics and steady-state response characteristics of these vehicle state parameters.

Simulation condition: on a road surface with an adhesion coefficient of 0.85, the vehicle is driven straight forward at a speed of 90km/h. After a period of 0.1s, the steering wheel angle is stepped by 30°, and the experimental process keeps the vehicle speed at the constant value.

From Figure 5.11(a) and Table 5.3, it can be seen that compared with the conventional FWS, the sideslip angle of the feed-forward proportional control is reduced, but the reduction is relatively small. Both the LQR active rear-wheel steering control and the LQR active four-wheel steering control compress the vehicle's sideslip angle to a larger extent, with the LQR active four-wheel steering control having better compression effects and smaller fluctuations. The reduction of the sideslip angle causes the longitudinal axis of the vehicle to align with the direction of the vehicle's center of mass velocity during cornering, limiting its lateral slip, avoiding vehicle instability, and ensuring the vehicle's good tracking ability.

From Figure 5.11(b) and Table 5.4, it can be seen that all three control methods have a certain reduction in the vehicle's yaw rate, improving the vehicle's steering

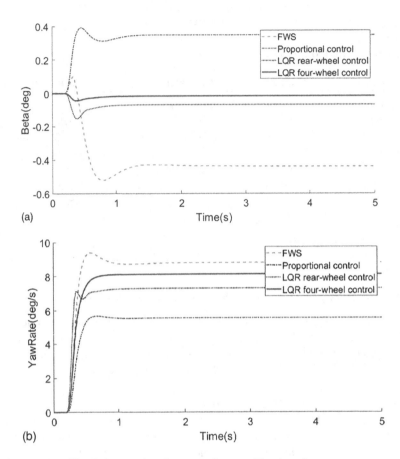

FIGURE 5.11　Simulation results of step steering test. (Continued)
Source: Figure created by authors.

stability at high speeds. Among them, the feed-forward proportional control significantly reduces the yaw rate, which can easily lead to a slow steering response. At the same time, it also forces the driver to turn more steering wheel angles compared to FWS vehicles, increasing the driving burden. The yaw rate response of both LQR active rear-wheel steering and LQR active four-wheel steering does not overshoot, resulting in smoother steering. Moreover, compared to conventional FWS, the stability time of the yaw rate is reduced and the response is faster.

According to the front wheel angle curve of the vehicle in Figure 5.11(c), LQR active four-wheel steering has the maximum front wheel angle, which can compensate for the decrease in steering sensitivity caused by the introduction of the same direction angle of the rear wheels. To some extent, the vehicle's yaw rate is also increased, preventing the deterioration of the vehicle's emergency obstacle avoidance performance.

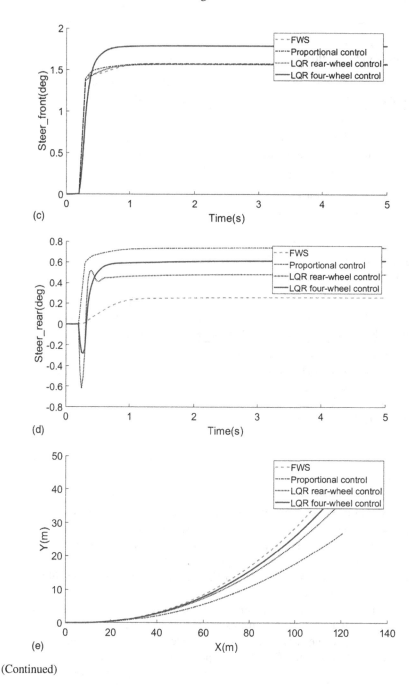

(Continued)

FIGURE 5.11 Simulation results of step steering test.

Source: Figure created by authors.

TABLE 5.3
Sideslip Angle Results

Control algorithm	Peak sideslip angle (deg)	Sideslip angle steady-state value (deg)
Conventional FWS	-0.521	-0.439
Feed-forward proportional control	0.391	0.347
LQR active rear-wheel steering control	-0.150	-0.067
LQR active four-wheel steering control	-0.043	-0.016

Source: Table created by authors.

TABLE 5.4
Yaw Rate Response Results

Control algorithm	Yaw rate steady state value (deg/s)	Yaw rate stabilization time (s)	Yaw rate overshoot
Conventional FWS	8.83	0.67	6.46%
Feed-forward proportional control	5.58	0.50	1.79%
LQR active rear-wheel steering control	7.32	0.52	0%
LQR active four-wheel steering control	8.14	0.55	0%

Source: Table created by authors.

According to the rear wheel angle curve of the vehicle in Figure 5.11(d), the rear wheel angles of LQR active rear-wheel steering and LQR active four-wheel steering have a reverse gain during the transient process, which can accelerate the vehicle's steering response and improve the vehicle's transient performance. Over time, the steering direction of the rear wheels becomes the same as that of the front wheels, suppressing the generation of the vehicle's sideslip angle.

From the vehicle trajectory curve in Figure 5.11(e), it can be seen that all three control methods have to some extent reduced the lateral displacement of the vehicle, increased the turning radius, and made the vehicle turn more stable at high speeds. However, the feed-forward proportional control causes the vehicle's steering response to be too slow, reducing the vehicle's emergency obstacle avoidance performance at high speeds. Relatively, the LQR control method has a smoother increase in turning radius.

5.3.2.2   Lateral Wind Interference Simulation Test

It is an ideal driving state for a vehicle to maintain a stable straight line without being affected by external factors. When vehicles are driving at high speeds, they are often disturbed by lateral winds, which can cause the tires to deviate at a certain angle and cause vehicle lateral displacement. At this time, the driver will unconsciously turn the steering wheel in the opposite direction to restore the vehicle to straight-line driving. However, due to excessive rotation of the steering wheel, there is a safety hazard of the vehicle slipping or even overturning, seriously affecting the handling and stability of the vehicle [3]. In this section, the simulated motion states of the vehicle based on the two LQR control methods will be investigated after the vehicle is disturbed by a steady lateral wind in a steady straight-line driving condition.

The lateral wind interference simulation conditions are set as follows: the vehicle is driving in a straight line at a speed of 90km/h, without turning the steering wheel, with a road adhesion coefficient of 0.85, and the lateral wind acting on the right side of the vehicle with a wind speed of 50km/h for 10 seconds.

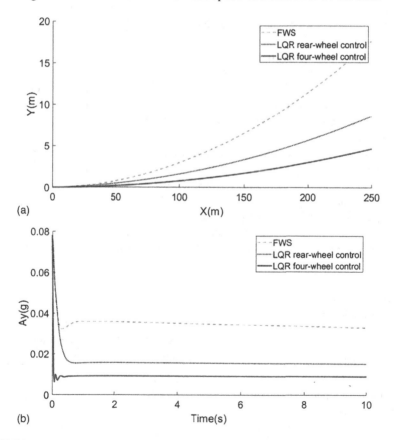

(a)

(b)

FIGURE 5.12   Simulation results of lateral wind interference test.

Source: Figure created by authors.

From the vehicle trajectory curve in Figure 5.12(a), it can be seen that, compared to traditional FWS, both LQR active rear-wheel steering control and LQR active four-wheel steering control can resist lateral deviation caused by lateral wind. Among them, LQR active four-wheel steering control is more optimal, improving the stability of the vehicle when driving in crosswind conditions.

From the lateral acceleration curve in Figure 5.12(b), it can be seen that compared to the traditional FWS lateral acceleration stable value of 0.033g, the LQR active rear-wheel steering control has a lateral acceleration stable value of 0.015g, and the LQR active four-wheel steering control has a lateral acceleration stable value of 0.009g, which enters the steady state faster.

5.3.2.3 Double Lane Change Simulation Test

In the double lane change test, the vehicle needs to travel from one lane to another and then return to the original lane. This working condition is mainly used to test the emergency obstacle avoidance ability of vehicles, and it is an important working condition for testing the performance of the steering system. The double lane change test is widely used in the vehicle industry [8].

(i) 120km/h double lane change test

Simulation condition: 120km/h vehicle speed, 0.85 road adhesion coefficient.

From the vehicle trajectory curves and lateral error curves in Figure 5.13(a) and (b), it can be seen that compared to the other three steering controls, LQR active four-wheel steering control can effectively complete double lane change test at high speeds, and has the best tracking effect for ideal trajectories. During the first lane-change process, the vehicle's response is the fastest, and the lateral overshoot is significantly smaller than that of FWS. During the second lane change back to the original driving road, it can enter faster and greatly reduce the occurrence of a tailspin. From Table 5.5, it can be seen that compared to conventional FWS, the average lateral error between LQR active four-wheel steering trajectory and ideal trajectory decreased from 0.1801m to 0.1292m, and the maximum value decreased

---

TABLE 5.5
Lateral Error Results

Control algorithm	Mean lateral error (m)	Max lateral error (m)
Conventional FWS	0.1801	1.24
Feed-forward proportional control	0.1838	1.29
LQR active rear-wheel steering control	0.1815	1.26
LQR active four-wheel steering control	0.1292	0.82

Source: Table created by authors.

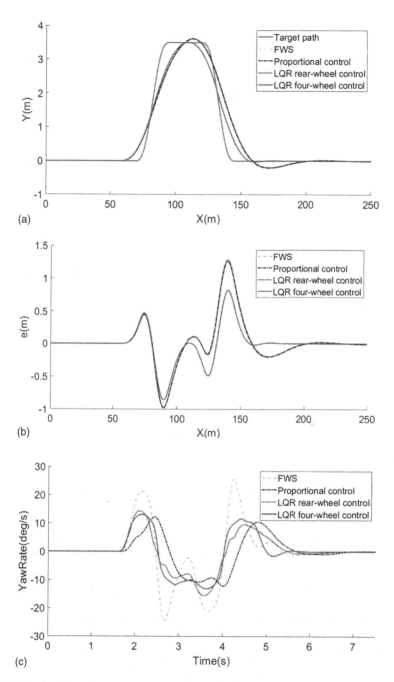

FIGURE 5.13 Simulation results of 120km/h DLC test. (Continued)
Source: Figure created by authors.

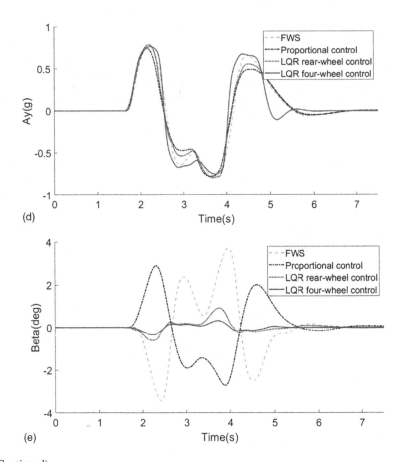

(Continued)

FIGURE 5.13   Simulation results of 120km/h DLC test.

Source: Figure created by authors.

from 1.24m to 0.82m, resulting in a smoother driving trajectory. The driving trajectories of the other three control strategies are not significantly different, so the control effects are similar.

From Figure 5.13(c) and (d), it can be seen that compared to conventional FWS, the LQR active four-wheel steering control significantly reduces the yaw rate of the vehicle in double lane change test, improving the comfort of drivers and passengers in the extreme situation of high-speed obstacle avoidance. At the same time, the lateral acceleration of the vehicle does not significantly decrease, maintaining the obstacle avoidance performance of the vehicle.

From the response curve of the sideslip angle in Figure 5.13(e), it can be seen that the maximum sideslip angles of the conventional FWS and three control strategies are 3.72°, 2.88°, 0.91°, and 0.32°, respectively. The LQR control strategy significantly reduces the sideslip angle in the double lane change test, and the LQR active four-wheel steering control has a significantly better compression effect on the sideslip angle than the LQR active rear-wheel steering control. The smaller sideslip angle ensures that the direction of the vehicle's driving speed almost coincides with the longitudinal axis of the vehicle body, ensuring a good posture during the steering process and improving the vehicle's steering safety.

(ii) 100km/h low road adhesion coefficient double lane change test

Simulation condition: 100km/h vehicle speed, 0.4 road adhesion coefficient. From the vehicle trajectory curves and lateral error curves in Figure 5.14(a) and (b), it can be seen that compared to the other three steering controls, LQR active four-wheel steering control has the best tracking effect on the ideal trajectory with high speed and on low adhesion roads. During the first lane change process, the lateral overshoot was significantly smaller than that of FWS. And during the second lane change back to the original driving road, it can enter faster and greatly reduce the occurrence of a tailspin. From Table 5.6, it can be seen that compared to conventional FWS, the average lateral error between LQR active four-wheel steering and the ideal trajectory decreased from 0.3694m to 0.3000m, and the maximum value decreased from 2.56m to 2.20m, resulting in a smoother driving trajectory. The driving trajectories of the other three control strategies are not significantly different, so the control effects are similar.

According to the response curve of the sideslip angle in Figure 5.14(c), the maximum sideslip angles of the conventional FWS and the three control strategies are 4.11°, 4.47°, 1.84°, and 1.64°, respectively. The LQR control strategy significantly reduces the sideslip angle of the vehicle in double lane change test on low adhesion roads, and the LQR active four-wheel steering control has a slightly better compression effect on the sideslip angle than the LQR active rear-wheel steering control.

Through the simulation experiments in this section, it can be seen that compared to conventional FWS vehicles, the LQR-controlled 4WS vehicle with the control objectives of tracking the ideal yaw rate and zero sideslip angle has improved the following vehicle performance. The optimized driving performance can be concluded as

(1) The turning radius of the vehicle is increased during the high-speed driving, and the steering stability of the vehicle can be improved at high speeds.

(2) The turning radius of the vehicle is reduced during the low-speed driving, and the vehicle's steering flexibility can be improved at low speeds.

(3) A faster steering response can be achieved while the steering transient overshoot is reduced.

(4) The sideslip angle of vehicle can be reduced during the steering control, and maintain better body posture and vehicle tracking performance.

(5) The overall stability of the vehicle during high-speed obstacle avoidance can be improved significantly.

(6) The stability of the vehicle is enhanced, and has ability to resist lateral wind interference, and to adapt for low adhesion coefficient road surfaces.

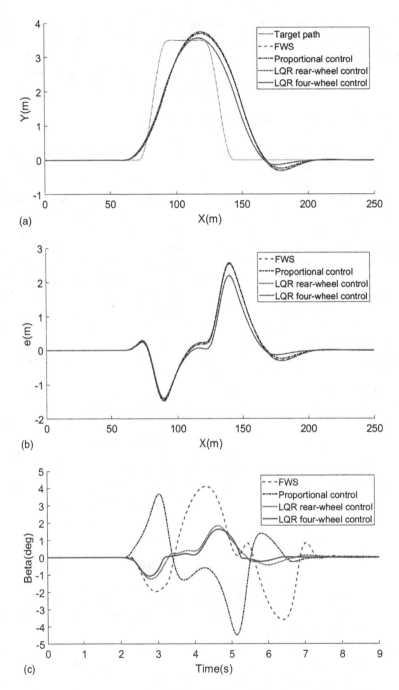

FIGURE 5.14   Simulation results of 100km/h 0.4 road adhesion coefficient DLC test.
Source: Figure created by authors.

TABLE 5.6
Lateral Error Results

Control algorithm	Mean lateral error (m)	Max lateral error (m)
Conventional FWS	0.3694	2.56
Feed-forward proportional control	0.3640	2.54
LQR active rear-wheel steering control	0.3727	2.56
LQR active four-wheel steering control	0.3000	2.20

Source: Table created by authors.

5.3.3 COMPARATIVE ANALYSIS OF ACTIVE FOUR-WHEEL STEERING CONTROL METHODS

In this section, we will conduct designed simulation experiments to compare the performance differences of three kinds of active four-wheel steering control strategies, which are LQR control, three-step control, and sliding mode control respectively.

5.3.3.1 Steering Wheel Angle Step Simulation Test

Simulation condition: on a road surface with an adhesion coefficient of 0.85, the vehicle is driven straight forward at a speed of 90km/h. After a period of 0.1s, the steering wheel angle is stepped by 30°, and the experimental process keeps the vehicle speed at the constant value.

From Figure 5.15(a) and Table 5.7 it can be seen that, compared to conventional FWS, the three active four-wheel steering control strategies can greatly compress the vehicle's sideslip angle, and the compression effect is almost equivalent. Their peak sideslip angles are not significantly different, but the stable sideslip angles for three-step control and sliding mode control are smaller compared to LQR control. This may be related to the selection of control parameters. The overall control effect between these three control strategies is similar.

From Figure 5.15(b) and Table 5.8, it can be seen that compared to conventional FWS, the yaw rate of the three active four-wheel steering control strategies has been reduced, and the steering stability of the vehicle has been improved at high speeds. These strategies reduce the stabilization time of the yaw rate, making the response more rapid. And they also reduce the overshoot of the yaw rate and improve the transient characteristics of the yaw rate. Among them, sliding mode control has the fastest response, and LQR control has the smallest overshoot.

5.3.3.2 Double Lane Change Simulation Test

Simulation condition: 120km/h vehicle speed, 0.85 road adhesion coefficient.

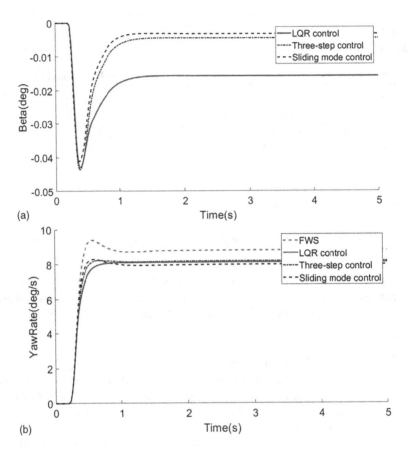

(a)

(b)

FIGURE 5.15   Simulation results of steering wheel angle step test.

Source: Figure created by authors.

TABLE 5.7
Sideslip Angle Results

Control algorithm	Peak sideslip angle (deg)	Sideslip angle steady-state value (deg)
Conventional FWS	-0.521	-0.439
LQR active four-wheel steering control	-0.043	-0.016
Three-step four-wheel steering control	-0.044	-0.005
Sliding mode four-wheel steering control	-0.041	-0.003

Source: Table created by authors.

TABLE 5.8
Yaw Rate Results

Control algorithm	Yaw rate steady state value (deg/s)	Yaw rate stabilization time (s)	Yaw rate overshoot
Conventional FWS	8.83	0.67	6.46%
LQR active four-wheel steering control	8.14	0.55	0%
Three-step four-wheel steering control	8.21	0.50	0.24%
Sliding mode four-wheel steering control	8.01	0.45	3.50%

Source: Table created by authors.

TABLE 5.9
Lateral Error Results

Control algorithm	Mean lateral error (m)	Max lateral error (m)
Conventional FWS	0.1801	1.24
LQR active four-wheel steering control	0.1292	0.82
Three-step four-wheel steering control	0.1198	0.66
Sliding mode four-wheel steering control	0.1348	0.87

Source: Table created by authors.

From the vehicle trajectory curves and lateral error curves in Figure 5.16(a) and (b), it can be seen that compared to conventional FWS, all three types of active four-wheel steering can effectively complete double lane change test at high speeds, and there will be no lateral overshoot after the second lane change. From Table 5.9, it can be seen that compared to conventional FWS, the trajectory lateral errors of these control strategies have significantly decreased, with the three-step control having the smallest lateral error value, followed by LQR control.

According to the response curve of the sideslip angle in Figure 5.16(c), the maximum sideslip angles of conventional FWS and three active four-wheel steering control strategies are 3.72°, 0.32°, 0.21°, and 0.18°, respectively. All three active four-wheel steering control strategies significantly reduce the sideslip angle of the vehicle in the double lane change test, and the control effect is similar.

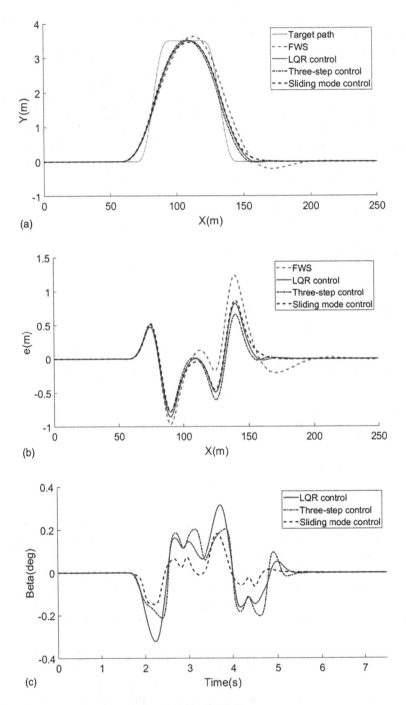

FIGURE 5.16 Simulation results of 120km/h DLC test.

Source: Figure created by authors.

Through the above simulation analysis, it can be seen that the three active four-wheel steering control strategies of LQR control, three-step control, and sliding mode control all improve the steering performance of the vehicle. Comparing these three active four-wheel steering control strategies, we can find that their control effects are not significantly different from each other, which is mainly related to the adjustment of control parameters and needs further refinement. The control parameters of the LQR control have a clear impact on steering performance (sideslip angle compression effect and ideal yaw rate tracking effect), making it easy to adjust the parameters.

REFERENCES

[1] Zhao, W., Zhang, H., Zou, S., et al. (2021). Overview of vehicle steer-by-wire system control technologies. *J. Automot. Saf. Energy*, 12(01):18–34.

[2] Sano, S., Furukawa, Y. and Shlralshi, S. (1986). Four wheel steering system with rear wheel steer angle controlled as a function of steering wheel angle. SAE Trans., 95: 880–893.

[3] Shibahata, Y., Shimada, K., Tomari, T. (1993). Improvement of vehicle maneuverability by direct yaw moment control. *Veh. Syst. Dyn.*, 22(5–6):465–481.

[4] Lu, A., Lu, Z., Li, R., et al. (2022). Adaptive LQR path tracking control for 4WS electric vehicles based on genetic algorithm[C]. *2022 6th CAA International Conference on Vehicular Control and Intelligence (CVCI)*, pages 1–6. IEEE.

[5] Chen, X., Han, Y., Hang, P. (2020). Researches on 4WIS-4WID stability with LQR coordinated 4WS and DYC, advances in dynamics of vehicles on roads and tracks. *Proceedings of the 26th Symposium of the International Association of Vehicle System Dynamics, IAVSD 2019*, August 12-16, Gothenburg, pages 1508–1516. Springer International Publishing, Sweden. Cham.

[6] Li, H., Li, P., Yang, L., et al. (2022). Safety research on stabilization of autonomous vehicles based on improved-LQR control. *AIP Adv.*, 12(1):015313.

[7] Norouzi, A., Adibi Asl, H., Kazemi, R., Hafshejani, P. F. (2020). Adaptive sliding mode control of a four-wheel-steering autonomous vehicle with uncertainty using parallel orientation and position control. *Int. J. Heavy Veh. Syst.*, 27(4).

[8] Yuhara, N., Horiuchi, S., Arato, Y. (1992). A robust adaptive rear wheel steering control system for handling improvement of four-wheel steering vehicles. *Veh. Syst. Dyn.*, 20(supl):666–680.

6 Variable Steering Ratio Control of Steer-by-Wire Vehicle

6.1 INTRODUCTION OF VARIABLE STEERING RATIO

As the main property of lateral dynamics, vehicle cornering stability and handling performance are greatly influenced by the steering system. The angular transmission ratio of an automotive steering system is usually defined as the ratio between the rotation angle of steering handwheel and vehicle front wheels. In conventional steering systems, the steering angle is transmitted by means of mechanical transmission. Therefore, once the mechanical structure is determined, the angular transmission ratio between the steering wheel and the front wheel is basically fixed. This makes the dynamic performance of conventional mechanical steering system is limited. In the low-speed steering condition (e.g., parking), the driver needs to frequently make a large steering angle, which increases the burden of the driver. On the contrary, a smaller angular ratio can improve the steering sensitivity of the vehicle, but it will increase the difficulty of controlling the vehicle at high speeds. Therefore, the traditional steering systems with fixed steering ratio have difficulty balancing the driving performance between different vehicle speeds.

To improve the driving experience, an adaptive steering characteristic called the variable gear-ratio steering (VGS) control was developed for vehicles, which has changeable steering gain with respect to the different states of vehicle dynamics. In previous studies, the researcher found that a larger steering wheel angle manipulation could increase driver's physical workload, and the meticulous steering operation could also increase driver's mental workload [1]. A typical VGS system is active front steering system from BMW corporation [2], which utilizes a double planetary gear system and an electric actuator motor to facilitate driver independent steering of the front wheels. This system is intended to improve the steering stability at different driving speed by an optimized transmission coefficient. Because of the mechanical decoupled structure of steer-by-wire (SBW) system, the steering control with variable transmission ratio can be conducted through a software, without any requirement of a hardware configuration.

The angular transmission of the SBW system can be divided into two parts. The first part of the mechanical transmission is from the rack and pinion steering gear

 DOI: 10.1201/9781003481669-6

to the vehicle front wheels. The second part is the angular transmission from the steering handwheel to the steering actuator. Due to physical decoupling of SBW, the second part are conducted by electronic transmission. The SBW controller can calculate the target steering angle with variable transmission ratio. The rack and pinion steering gears generate the desired angle by the control of steering actuator. For the convenience of calculation, the variable transmission ratio of SBW is defined as the ratio between the steering wheel angle and the target pinion angle, which is expressed as:

$$\beta = \frac{\theta_s}{\theta_p} \tag{6.1}$$

where β is the SBW angular ratio, θ_s and θ_p are the steering wheel angle and steering pinion angle, respectively. The variable steering ratio of SBW system is shown in Figure 6.1.

The control method of variable steering ratio for the SBW system will be introduced in this chapter. In order to improve the vehicle handling performance, the variable steering ratio characteristic of the SBW vehicle is designed by considering not only vehicle steering response, but also the driver's steering workload. The optimal steering ratio characteristics with respect to different driving cycles are evaluated by the driver–vehicle simulation system. Moreover, a stabilized yaw rate gain is considered in the design of steering ratio to achieve linear steering response. In the following sections, the performance of variable steering ratio with different vehicle driving states are discussed.

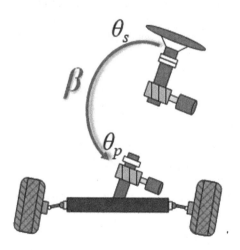

FIGURE 6.1 Variable steering ratio of steer-by-wire system.
Source: Figure created by authors.

6.2   STEERING RATIO CHARACTERISTICS DEPENDENT ON VEHICLE SPEED

During the low speed or stationary steering, the steering sensitivity is the main optimization goals for the SBW vehicle. The SBW system is required to have a smaller angular transmission ratio to satisfy the requirements of steering sensitivity, and also reduce the driver's control burden. With the increase of vehicle speed, it is expected to have a larger steering angular transmission ratio, in order to avoid the vehicle drifting due to high steering sensitivity. Meanwhile, a larger steering angular transmission ratio will reduce the control response capabilities, and cannot meet the requirements of emergency driving behaviors, such as emergency obstacle avoidance. In order to improve the control response and stability for different driving conditions, the SBW angular transmission ratio should be designed with the consideration of vehicle velocity, which is shown as:

$$\beta = \begin{cases} \beta_{min} & v_x \leq v_0 \\ f(v_x) & v_0 < v_x < v_1 \\ \beta_{max} & v_x \geq v_1 \end{cases} \qquad (6.2)$$

Where β_{min} is the lower limit value of angular transmission ratio, v_0 is the lower limit value of angular transmission ratio corresponding to the vehicle speed (km/h), $f(v_x)$ is the nonlinear characteristic of angular transmission ratio change with vehicle speed, v_1 is the upper limit value of angular transmission ratio corresponding to the vehicle speed (km/h), β_{max} is upper limit value of angular transmission ratio.

Considering that the performance of vehicle steering control will be affected by many factors of vehicle dynamic, this chapter introduces a kind of objective evaluation index of vehicle maneuvering performance. Based on this objective evaluation index, the optimal mapping relationship between the angular transmission ratio and vehicle speed is determined through the analysis of double lane change (DLC) experiments.

6.2.1   OBJECTIVE EVALUATION INDICATORS OF VEHICLE MANEUVERING PERFORMANCE

To determine the relationship between SBW angular transmission ratio β and vehicle speed v_x, a multi-objective evaluation index system for vehicle handling performance is designed. Based on the vehicle dynamic states in Figure 6.2, the evaluation models are expressed by the following quadratic cost functions.

(1) Indicators for considering vehicle path-following errors J_e:

$$J_e = \int_0^t \left(\frac{e}{e_t} \right)^2 dt \qquad (6.3)$$

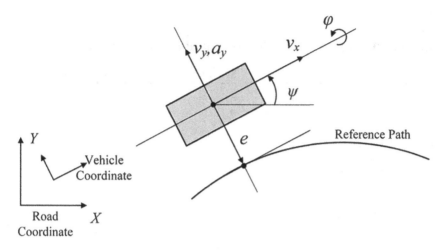

FIGURE 6.2 Sketch of the vehicle dynamic states used for the evaluation index.
Source: Figure created by authors.

where t is experimental time (same as below), e is lateral displacement deviation (m) between the actual traveling path of the vehicle and the desired path, e_t is standard threshold of lateral displacement deviation (m).

(2) Indicators that consider the steering burden on the driver J_b:

$$J_b = \int_0^t \left(\frac{\dot{\theta}_s}{\dot{\theta}_{st}} \right)^2 dt \qquad (6.4)$$

where $\dot{\theta}_s$ is steering wheel rotation angular velocity (rad/s), $\dot{\theta}_{st}$ is standard threshold of steering wheel rotation angular velocity (rad/s).

(3) Indicator J_c, which considers the lateral acceleration of the vehicle:

$$J_c = \int_0^t \left(\frac{a_y}{a_{yt}} \right)^2 dt \qquad (6.5)$$

Where a_y is vehicle lateral acceleration (g), a_{yt} is vehicle lateral acceleration standard threshold (g).

(4) Indicators to consider the risk of vehicle rollover J_r:

TABLE 6.1
Threshold Values Used for Objective Evaluation Index

Variable	Symbol	Standard Threshold	Unit
Lateral displacement deviation	e_t	0.4	m
Angular speed of steering wheel rotation	$\dot{\theta}_{st}$	1.0	rad/s
Lateral acceleration	a_{yt}	0.3	g
Vehicle roll angle	φ_t	3.0	°

Source: Table created by authors.

$$J_r = \int_0^t \left(\frac{\varphi}{\varphi_t} \right)^2 dt \tag{6.6}$$

Where φ is vehicle lateral inclination angle (°), φ_t is standardized threshold value of vehicle lateral inclination angle (°).

After weighting and combining the above four sub-indicators, a comprehensive evaluation index J is obtained as shown in the following equation. The smaller the index, the better the performance of the automobile's maneuvering stability.

$$J = \sqrt{\frac{J_e^2 + J_b^2 + J_c^2 + J_r^2}{4}} \tag{6.7}$$

The weighting coefficient value of each indicator in the above equation can be selected by adjusting the standard threshold value [3]. Based on the driver-vehicle simulation system, the corresponding standard threshold values of each variable of the objective evaluation indexes are shown in Table 6.1.

6.2.2 DOUBLE LANE CHANGE EXPERIMENT AND ANALYSIS

Based on the above objective evaluation indexes, the results of normalized performance index with different compensating coefficient of steering ratio β are investigated. According to ISO standard for steering experiment [4], a double lane change (DLC) test condition is employed in the simulation with specified vehicle speeds. The design of driving path of DLC is shown in Figure 6.3. In the experiment, six different desired speeds are set as 20, 40, 60, 80, 100 and 120km/h. The SBW vehicle with different angular transmission ratios β will take double lane change, and the values of objective evaluation indexes are calculated at the same time.

Figure 6.4 shows the comparison results of the objective evaluation indexes for DLC experiment with different angular ratios β at different vehicle speeds. From

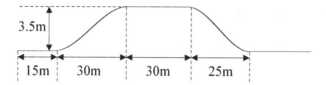

FIGURE 6.3   Reference driving path designed for DLC task.
Source: Figure created by authors.

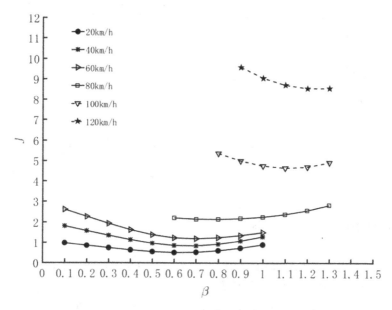

FIGURE 6.4   Results of DLC experiments with specified vehicle speeds.
Source: Figure created by authors.

the results, at a given speed, the index J decreases and then increases with the increase of β. It means there is an optimal point for β corresponding to the vehicle speed. For example, from the date curve of the speed of 100 km/h, the optimal ratio is $\beta = 1.1$, where a smallest value of J is achieved.

Based on the above simulation results and analysis, the optimal SBW angular ratios corresponding to given vehicle speeds are determined. By using data fitting, the optimal SBW angular ratio characteristics with vehicle speed as shown in Figure 6.5. Compared with the conventional fixed steering system ($\beta = 1$), the variable ratio of the SBW system show a nonlinear growth with the change of vehicle speed. In addition, the angular ratio β satisfies the upper and lower limits of Equation (6.2), with β_{max} and β_{min} being 1.3 and 0.6, respectively.

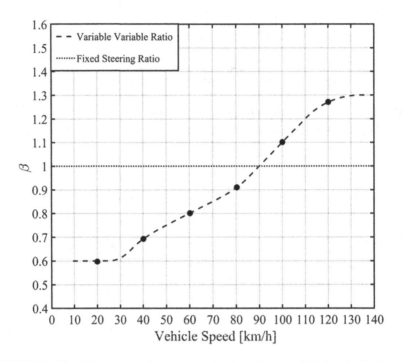

FIGURE 6.5   Variable steering ratio characteristics dependent on vehicle longitudinal speed. Source: Figure created by authors.

6.3   STEERING RATIO CHARACTERISTIC DEPENDENT ON STEERING WHEEL ANGLE

Based on the vehicle handling dynamics, the yaw rate gain during the steering control is an important parameter to show the handling performance of vehicle [5]. The yaw rate gain G_s^γ is used as an indicator of vehicle steering characteristics. It is calculated by the steering wheel angle θ_s and the vehicle yaw rate response γ, which is expressed as:

$$G_s^\gamma = \frac{\gamma}{\theta_s} \tag{6.8}$$

By conducting steering angle step input experiments, the transient steering characteristics of the vehicle is analyzed. The graphs of the steady state yaw rate gain G_s^γ versus the steering wheel angle at different vehicle speeds are obtained in Figure 6.6. And the relationship curves between yaw rate γ and the steering pinion angle θ_p for various speeds of the vehicle are obtained in Figure 6.7. From the results of two figures, it can be found that the response of the vehicle's yaw rate shows a linear relation with the pinion angle when the steering angle is small. The steering gain G_s^γ is basically a constant value in small steering angle. However, as the angle increases, the vehicle tire begins to enter the nonlinear region, and the steering yaw rate gain begins to show a gradually decreasing characteristic. This kind of nonlinear response characteristic will force the driver to continuously adjust the steering gain to maintain a desired vehicle steering performance, which greatly increases the physical and mental burden of the driver. In order to compensate for the decreasing characteristic of yaw rate gain, a variable steering ratio can be designed to change with the steering wheel angle, and reducing the driver's maneuvering burden under different operating conditions.

From Figure 6.6, each curve corresponding to a vehicle speed has a basically constant yaw rate gain in the region of small steering angle. This constant value is the ideal steering gain $(G_s^{\gamma})_{des}$ at this vehicle speed. The ideal yaw rate γ_{des} of the steering input can be calculated by the following equation:

$$\gamma_{des} = (G_s^{\gamma})_{des} \cdot \theta_s \qquad (6.9)$$

The relationship curves between yaw rate γ and the steering pinion angle θ_p for various speeds of the vehicle are shown in Figure 6.7. If the vehicle speed is constant, the, there is an ideal pinion angle $\left(\theta_p\right)_{des}$ corresponding to the steering wheel angle. Based on the equation 6.9, the compensating coefficient of the steering ratio is designed by the following equation

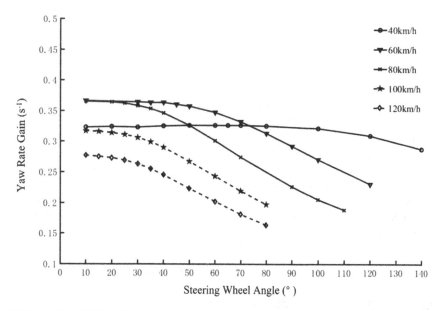

FIGURE 6.6   SBW vehicle yaw rate gain properties of with different longitudinal speeds.
Source: Figure created by authors.

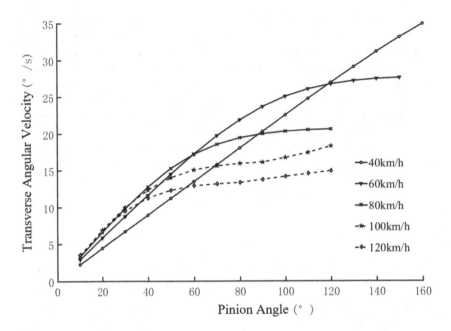

FIGURE 6.7   SBW vehicle yaw rate response to steering pinion angle with different longitudinal speeds.

Source: Figure created by authors.

$$\beta_{des} = \frac{\theta_s}{\left(\theta_p\right)_{des}} \qquad (6.10)$$

With the increase of steering angle, the tire will gradually transition from the linear characteristic region to the nonlinear characteristic region. In order to ensure the stability of the vehicle handling response, the steering ratio should be kept as constant as possible. Finally, the characteristics of the variable steering angle ratio corresponding to the steering wheel angle and vehicle longitudinal speeds are shown in Figure 6.8.

Based on the results of optimized steering ratio characteristic in Figure 6.8, the angle transmission of SBW can be controlled by both of steering angle input and corresponding vehicle longitudinal speeds. By the compensation control of the steering ratio, the handling performance of SBW vehicle can be improved by a relatively linear vehicle yaw rate.

6.4   CONTROL OF VARIABLE STEERING ANGLE RATIO

6.4.1   Design of Fuzzy Neural Networks

Fuzzy neural network (FNN) is a control method by the combination of fuzzy theory and neural networks, and this kind of hybrid control method was first

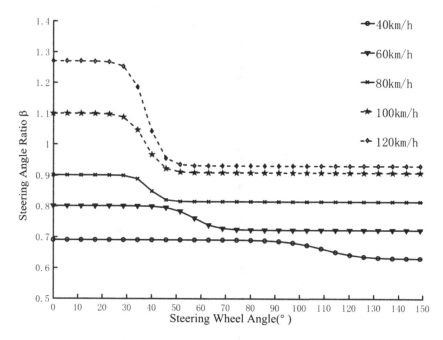

FIGURE 6.8 Variable steering angle ratio characteristics dependent on steering wheel angle for specific longitudinal speeds.

Source: Figure created by authors.

proposed by Lee S C and Lee E T in 1974 [6]. Fuzzy theory is an empirical approach which does not rely on precise mathematical models. It realizes desired control through an artificially established fuzzy rule base, and achieves knowledge reasoning by using the matching of rules. Fuzzy-based control has the advantages of fast processing speed and strong reasoning ability. However, the disadvantage is that the performance depends heavily on the experiences of the designer. To compensate for the shortcomings of Fuzzy-based control, the neural network is employed to generate the self-learning and self-adaptation abilities. Therefore, fuzzy neural network is an optimization of fuzzy control theory, which has the advantages of fuzzy theory's good use of empirical knowledge and strong reasoning ability, and has the advantages of neural network's good knowledge from data and strong learning ability.

In this section, a fuzzy neural network is used to design the variable steering angle ratio controller for SBW vehicles. Based on the above simulation results of variable steering ratio characteristic dependent on vehicle longitudinal speed and steering angle, it is a complex nonlinear system with MISO (multiple input and single output). To solve this kind of nonlinear control issue, a fuzzy neural network based on Takagi-Sugeno model is adopted in this section. The simulation data of optimized steering ratio are used to train the fuzzy neural network controller. The structure of the proposed fuzzy neural network is illustrated in Figure 6.9, and the design of five layers is described as follows:

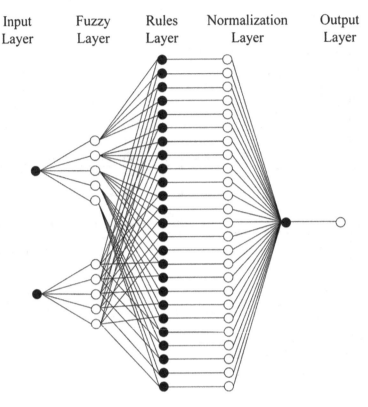

FIGURE 6.9   Structure of proposed T-S fuzzy neural network model.

Source: Figure created by authors.

1. Input layer: Vehicle longitudinal speed and steering wheel angle are the input variables of steering ratio controller. The role of this layer is to pass the input variables to the next layer. The input node is set as x_i, and output node is set as:

$$O_i^{(1)} = x_i, i = 1, 2 \qquad (6.11)$$

2. Fuzzy Layer: The role of this layer is to fuzzify the input variables separately to find out the degree of affiliation of each input. Five fuzzy sets with Gaussian membership functions are designed, that is, very small (VS), small (S), medium (M), big (B) and very big (VB). The input node in this layer is from the results of equation 6.11, and the output node is:

$$O_{ij}^{(2)} = \mu_{ij}(x_i) = \exp\left(\frac{-(x_i - c_{ij})^2}{d_{ij}^2}\right), j = 1, 2, ..., 5 \qquad (6.12)$$

where $\mu_{ij}(x_j)$ denotes the value of the affiliation function of variable x_j to the jth linguistic variable. c_{ij} and d_{ij} denote the centre and width of the affiliation function respectively, which are the parameters to be trained.

3. Fuzzy Rule Layer: Each node in this layer represents a fuzzy inference rule, and there are totally 25 nodes. After the calculation of this layer, the fitness α_m of each rule can be obtained. The output of this layer is:

$$
\begin{aligned}
O_m^{(3)} &= \alpha_m = \mu_{1j_1}(x_1) \times \mu_{2j_2}(x_2), \\
j_1, j_2 &= 1, 2, \dots, 5; m = 1, 2, \dots, 25
\end{aligned}
\tag{6.13}
$$

4. Normalization layer: The role of this layer is to normalize the fitness of each rule, the number of nodes is the same as the third layer, and its output is expressed as:

$$
O_m^{(4)} = \overline{\alpha_m} = \frac{\alpha_m}{\sum\limits_{k=1}^{25} \alpha_k}
\tag{6.14}
$$

5. Output layer: The final output of this layer is the angular transmission ratio β, which is the weighted sum of the adaptations of each rule:

$$
\beta = O^{(5)} = \sum_{k=1}^{25} \rho_k \times \overline{\alpha_k}
\tag{6.15}
$$

where ρ_m is the connection weight coefficient for each rule adaptation.

6.4.2 PARAMETERS LEARNING AND NETWORK TRAINING

After determining the structure of the fuzzy neural network, it is necessary to choose an appropriate neural network algorithm for the parameters learning. The T-S type fuzzy neural network structure established in this section is a multi-layer feed-forward neural network, so the back-propagation algorithm is used for parameter learning and adjustment. The back-propagation algorithm mainly consists of two processes: forward propagation of information and backward propagation of error [7].

When a set of samples of capacity n is fed into the fuzzy neural network, the network produces the corresponding actual output. During the training of fuzzy neural network, an evaluation function is designed to show the error between the actual output and desired output:

$$
E = \frac{1}{2} \sum_{i}^{n} (y_{di} - y_{ti})^2
\tag{6.16}
$$

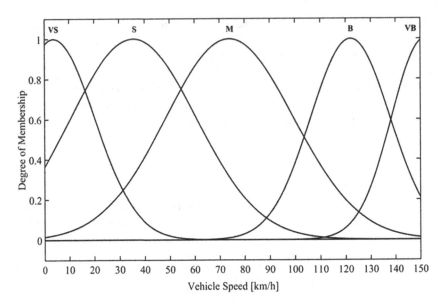

FIGURE 6.10 Membership function for vehicle longitudinal speed.

Source: Figure created by authors.

Based on the gradient descent method, the tuning algorithm for each network parameter is expressed as:

$$c_{ij}(k+1) = c_{ij}(k) - \eta \frac{\partial E}{\partial c_{ij}}$$

$$d_{ij}(k+1) = d_{ij}(k) - \eta \frac{\partial E}{\partial d_{ij}} \qquad (6.17)$$

$$\rho_m(k+1) = \rho_m(k) - \eta \frac{\partial E}{\partial \rho_m}$$

where $\eta > 0$ is the learning rate.

The sample data of variable steering ratio from the above simulation results are used to train the network. The number of iterations is set to 50,000, and the iteration is terminated when the error is less than 0.001. Figure 6.10 and 6.11 show the trained membership functions for the vehicle longitudinal speed and steering angle, respectively.

The output surface of the trained fuzzy neural network controller is shown in Figure 6.12. It characterizes the variable steering angle ratio of the SBW system as a function of vehicle speed and steering wheel angle. From the results, the control principles can be concluded as:

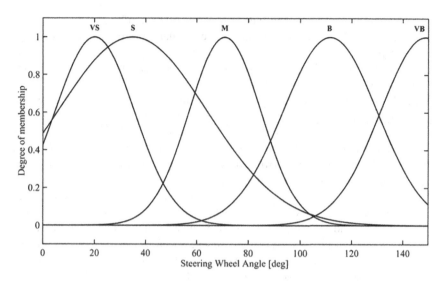

FIGURE 6.11   Membership function for steering wheel angle.
Source: Figure created by authors.

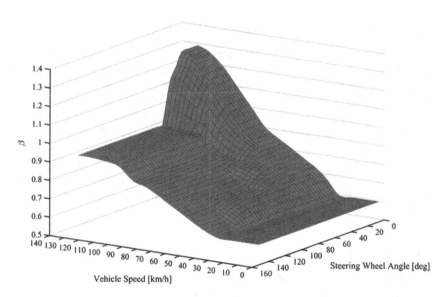

FIGURE 6.12   Fuzzy neural network inference system output surface.
Source: Figure created by authors.

1. If the vehicle is driving with a low speed (<30km/h), the steering angle ratio β will take the lower limit of the angular ratio. In this kind of driving condition, a small steering wheel angle can be used to obtain a large front wheel angle. The vehicle with good steering sensitivity can improve the low-speed steering performance.

2. If the vehicle gradually increases to the high speed (> 90km/h), the steering angle ratio β gradually increases to 1. The vehicle has quick response characteristics to obtain desired yaw and lateral control, and the yaw rate gain is maintained in a linear range by variable steering angle ratio. As the vehicle speed continues to increase, β will continue to increase to the upper limit of the angular ratio, β_{max} =1.3.

3. If the vehicle is steering with a constant longitudinal speed, the steering angle ratio β decreases as the steering wheel angle increases. The main objective is to expand the range of the linear response of the vehicle yaw rate to the steering wheel angle. When the vehicle speed is low, β decreases slightly corresponding to a large angle steering input. When the vehicle speed is high, β decreases rapidly corresponding to medium angle steering input. When the vehicle tire dynamics is about to enter the unstable nonlinear area, β remains unchanged to ensure that the vehicle has a relatively stable response.

6.5 COMPARISON EXPERIMENT AND ANALYSIS

In this section, the control of variable angular ratio by the proposed fuzzy neural network controller is verified by a driver-vehicle experiment system. The closed-loop driver-vehicle system is designed for the steering control dynamic simulation. The driver model controller algorithm is based on a standard single-track vehicle model. An optimal preview control theory from MacAdam[8] is applied to the driver-vehicle controller. According to the information of the predefined road model, the current position of the vehicle and the reference lateral displacement can be calculated. By using the single-track vehicle model, an expected steering angle is generated from the optimal preview control. The algorithm is embedded into the SBW vehicle with dynamics model, and a fixed angular ratio (β =1) steering system is selected as a comparison [9].

6.5.1 LEMNISCATE CURVE TEST

The Lemniscate curve tests are conducted to show the influence of the variable-angle ratio on the driver maneuvering burden. The test vehicle is controlled in the process of low and medium-speed driving, and its reference route is set up as shown in Figure 6.13. The vehicle speeds of 20km/h and 40km/h are selected as the experimental speeds. The SBW vehicle is controlled to take steering motion with two modes: variable-angle ratio mode and fixed-angle ratio mode. The driving control for one lap is used to evaluate the steering performance.

The Lemniscate curve tests have frequent large angle steering operations, which will increase the driver's physical and mental burden. Figure 6.14(a) and

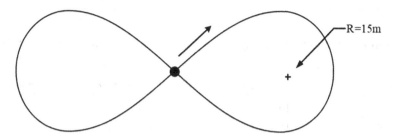

FIGURE 6.13 Driving route of lemniscate curve test.

Source: Figure created by authors.

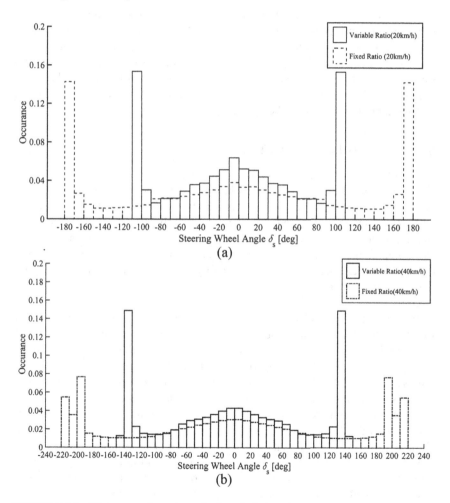

FIGURE 6.14 Histogram of steering wheel angle in lemniscate curve tests:

(a) $v_x = 20$ km/h and (b) $v_x = 40$ km/h.

Source: Figure created by authors.

TABLE 6.2
Maximum Steering Wheel Angle Input Summary for Lemniscate Curve Tests

Max. steering angle (deg)	Variable steering ratio	Fixed steering ratio	Ratio
20 km/h	107.61	174.03	0.62
40 km/h	140.71	218.27	0.64

Source: Table created by authors.

6.14(b) show the statistical results of the steering data distribution of the two kinds of steering modes. It can be found that the probability of occurrence for a large steering angle is reduced significantly by the variable steering ratio control. From the statistical results in Table 6.2, the maximum steering angle of variable steering ratio system reduced by approximately 40% than that of fixed steering ratio system. The control of steering ratio for the SBW system has optimized the steering amplitude to reduce the operating loads of drivers.

6.5.2 DOUBLE LANE CHANGE TEST

The double lane change (DLC) test is a typical steering condition. It is mainly used to evaluate the vehicle's path tracking performance and emergency obstacle avoidance performance at medium and high speeds. The reference route of DLC test is shown in Figure 6.16. The results of the experiments with variable steering ratio and fixed steering ratio are selected for comparison and analysis. The comparison of vehicle trajectories is shown in Figure 6.16, and the results of steering wheel angle and angular velocity are shown in Figure 6.16. For the purposes of comparison, the vehicle in the DLC test is driving with a constant 80km/h.

In Figure 6.15, the black solid line represents the predefined reference path for DLC test. The dash line and dash-dotted line are the driving paths of SBW vehicle with variable steering ratio system and fixed ratio system, respectively. Since the predictive control properties of vehicle driver model cannot accurately represent the real dynamics, there are increased tracking errors at the beginning of the steering control. The steering wheel angle and steering angular speed during the DLC test are showed in Figure 6.16. The evaluation index of steering control with fixed steering ratio and with variable steering ratio is compared in Table 6.3. It can be found that the lateral tracking error at turning points A and B is improved by the SBW system with variable ratio control. By using the normalized performance index in equation (6.7), the result of variable steering ratio has a smaller value than that of fixed steering ratio, which means it has a better handling performance.

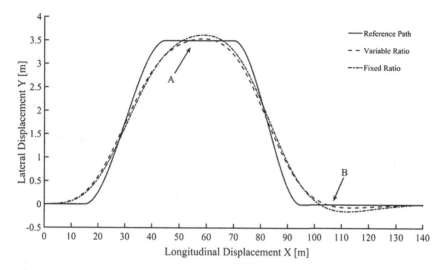

FIGURE 6.15   Comparison of driving trajectories of the DLC tests.
Source: Figure created by authors.

6.5.3   STEADY-STATE CIRCULAR TEST

The main objective of the steady state circular test is to evaluate the understeer and oversteer characteristics of the vehicle. From the previous study of steering system[10], the driver's workload will be increased when driving a vehicle with oversteer characteristics. Therefore, it is usually expected that the vehicle has understeer characteristics from the point of view of driving controllability and comfortable.

The steady state circular test is conducted with a fixed steering wheel angle, and continuous vehicle longitudinal acceleration control. The vehicle is controlled to track a circular route, which has a radius R_0 =15m. In the initial stages, the vehicle was accelerated to a stable state with a lateral acceleration of 0.1g. By keeping the steering handwheel with a constant angle, the vehicle was gradually accelerated until the lateral acceleration of the vehicle reached 0.7g. It is noted that the longitudinal acceleration cannot exceed 0.25g.

Figure 6.17 shows the comparison of the vehicle trajectories by the steering control with variable steering ratio and fixed steering ratio respectively. Figure 6.18 shows the characteristic curve of the steering radius ratio with lateral acceleration. As the longitudinal speed increases, the steering radius increases, indicating that both steering systems have certain understeering characteristics. From the results, the variable steering angle ratio control has a better handling performance due to the relative stable turning radius. Moreover, the adaptive steering ratio does not change the steering characteristics or controllability of the vehicle.

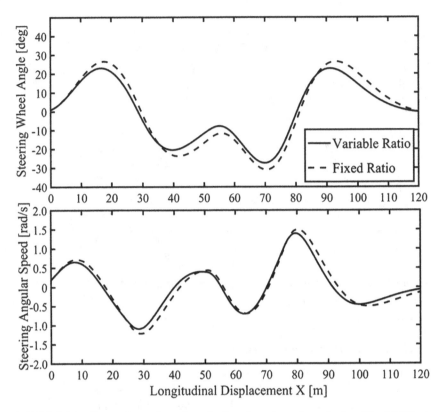

FIGURE 6.16   Comparison of steering wheel angle and angular velocity of the DLC tests.
Source: Figure created by authors.

TABLE 6.3
Evaluation Index Values of the DLC Tests

Evaluation index	Variable steering ratio	Fixed steering ratio
Steering angle range (deg)	23.01/-27.39	26.53/-30.94
Overshoot at point A (m)	0.038	0.118
Overshoot at point B (m)	0.062	0.147
Index *J*	1.59	2.06

Source: Table created by authors.

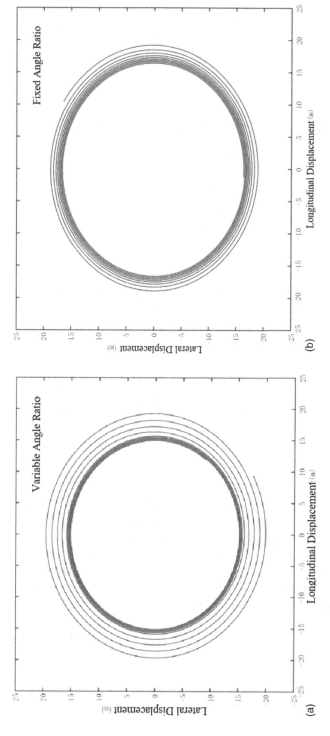

FIGURE 6.17   Comparison of vehicle driving trajectory during steady state circular tests.

Source: Figure created by authors.

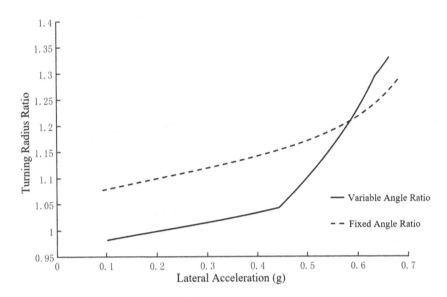

FIGURE 6.18 Turning radius ratio characteristics during steady state circular tests. Source: Figure created by authors.

6.6 CONCLUSION

By utilizing the characteristic of the steer-by-wire system, a variable steering ratio characteristic for steer-by-wire system is designed to improve vehicle handling performance. The steering ratio is adjusted by a compensating coefficient according to vehicle longitudinal speed and steering wheel angle. To evaluate the performance of vehicle with variable steering ratio, simulations are conducted based on an objective evaluation index, which consists of quadratic cost functions of vehicle lateral deviation, steering angular speed, vehicle lateral acceleration, and roll angle.

The control of variable steering ratio for SBW vehicle is designed by on a fuzzy neural network. Based on a closed-loop driver-vehicle system, a multi-objective evaluation method is designed by the quadratic cost functions of vehicle dynamic states. To achieve a desired steering ratio, a five-layer Takagi-Sugeno type fuzzy neural network is employed as the nonlinear controller. The evaluated results with an optimized steering ratio are used to train the network parameters. To verify the proposed control strategy, comparative experiments of lemniscate curve tests and DLC tests are conducted with the variable steering ratio and fixed steering ratio, respectively. The results show that the proposed SBW controller can not only improve handling performance at different longitudinal vehicle speed, but also reduce the steering burden of driver.

REFERENCES

[1]   Shimizu Y, Kawai T and Yuzuriha J. (1999) Improvement in driver-vehicle system performance by varying steering gain with vehicle speed and steering angle: VGS. *SAE technical paper* 1999-01-0395.

[2]   Kumar EA and Kamble DN. (2012) An overview of active front steering system. *International Journal of Scientific & Engineering Research*, 6: 1–10.

[3]   Zong Changfu, Guo Konghui. (2000) Objective quantitative evaluation indexes of automobile handling stability. *Journal of Natural Science of Jilin University*, 30(1):1–6.

[4]   ISO 3888-1:1999 Passenger cars-test track for a severe lane change manoeuver Part 1: double lane-change. 1999.

[5]   Heathershaw A. (2004) Matching of chassis and variable ratio steering characteristics to improve high speed stability, SAE *World Congress, SAE*, 2004-01-1103.

[6]   LEE S C, LEE E T. (1974) Fuzzy sets and neural networks. *Journal of Cybernetics*, 4(2): 83–103.

[7]   Qiao Junfei, Han Honggui. (2013) Feedforward neural network analysis and design [M]. *Science Press*: 80–94.

[8]   MacAdam CC. (1981) Application of an optimal preview control for simulation of closed-loop automobile driving. *IEEE Trans Syst Man Cybernet*, 11: 393–399.

[9]   Xiaodong Wu, Wenqi Li. (2020) Variable steering ratio control of steer-by-wire vehicle to improve handling performance, *Proceedings of the Institution of Mechanical Engineers, Part D: Journal Automobile Engineering*, 234(2–3): 774–782.

[10] Yasube, Masato (Author), Yu, Fan (Translator). (2016) Vehicle maneuvering dynamics: theory and application. *Beijing: Machinery Industry Press*: 244–246.

Index

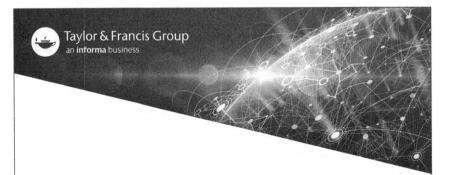

.

Printed in the United States
by Baker & Taylor Publisher Services